教育部“国家地质学理科基地双语教学系列课程建设”项目
和“国家地质学理科基地基金”项目联合资助

高等学校地质地矿类课程双语教学工具书

汉英 英汉
地球物质科学术语

Chinese-English English-Chinese
Glossary of Terms in Earth Material Science

主编　桑隆康　赵珊茸　李方林
编委　桑隆康　赵珊茸　李方林　肖 龙　张瑞生

中国地质大学出版社
ZHONGGUO DIZHI DAXUE CHUBANSHE

内容简介

本书包括结晶学和矿物学、晶体光学和光性矿物学、岩石学、矿床学等地球物质科学和地球科学概论英汉术语。为了保证基础性和先进性，收录了地球物质科学类课程基本专业术语，特别是近年出现的新词，除参考现有地质专业辞书外，以20世纪90年代以来国内外出版的相关新教材为基础进行编写工作。汉英地球物质科学术语部分包括6 221个词条，按汉语拼音字母顺序编排。英汉地球物质科学术语部分包括6 492个词条，按英文字母顺序排列。附录包括化学元素代号、矿物代号和国际地层表。

图书在版编目(CIP)数据

汉英英汉地球物质科学术语/桑隆康，赵珊茸，李方林主编. —武汉：中国地质大学出版社，2008.5

ISBN 978-7-5625-2257-7

Ⅰ.汉…

Ⅱ.①桑…②赵…③李…

Ⅲ.地球物理学-名词术语-汉、英

Ⅳ.P3-61

中国版本图书馆CIP数据核字(2008)第069836号

汉英英汉地球物质科学术语　　桑隆康　赵珊茸　李方林　主编

责任编辑：谌福兴　赵颖弘　　责任校对：戴　莹

出版发行：中国地质大学出版社(武汉市洪山区鲁磨路388号)　邮政编码：430074

电话：(027)67883580　传真：67883580　E-mail：cbb @ cug.edu.cn

经　销：全国新华书店　http://www.cugp.cn

开本：850毫米×1168毫米 1/32　字数：460千字　印张：13.25

版次：2008年5月第1版　印次：2008年5月第1次印刷

印刷：湖北地矿印业有限公司　印数：1—1 000册

ISBN 978-7-5625-2257-7　定价：78.00元

前　言

INTRODUCTION

当今世界科技发展日新月异，知识积累、更新和传播速度加快。而新知识国际交流主要载体和媒介如国际互联网、国际学术会议、科技刊物，无不使用英语。以地球科学和资源、环境科学刊物为例，世界200多种SCI检索系统源刊物全为英文版刊物。因此，英语已成为名副其实的国际科技交流语言，而英语交流能力当然也就成为大学理科学生素质乃至教员素质的重要方面。而大学理科学生的专业英语交流能力低是世界各个非英语的国家和地区普遍问题，这引起包括我国在内的世界各个非英语的国家和地区教育部门的高度重视。20世纪90年代以来，我国教育主管部门采取了大力强化增大英语课程比例、改善英语教学设施、聘请英、美籍英语教师、派遣理科留学生和访问学者出国、加强国际科技合作和交流、大力提倡“双语教学”等措施。目前，“双语教学”也已在地质地矿类专业中蓬勃开展。但由于缺乏合适的工具书，特别是近30年来地球物质科学迅速发展，出现了大量一般专业词典查不到的新词，使得学生在学习中遇到很大困难。为地质地矿类课程双语教学和全英语教学服务，给学生提供学习这些课程的工具书，同时也供地质地矿工作者参考，是这本《汉英英汉地球物质科学术语》的编写宗旨。

本“术语”包括结晶学和矿物学、晶体光学和光性矿物学、岩石学、矿床学等地球物质科学和地球科学概论英汉术语。为了保证基础性和先进性，收录了地球物质科学类课程基本专业术语，特别是近年出现的新词，除参考现有地质专业辞书外，以20世纪

90 年代以来国内外出版的相关新教材为基础进行编写工作。词条分工是：地球科学概论、晶体光学和光性矿物学、变质岩石学、矿床地质学、国家和地区、美国和加拿大行政区划名称、颜色等词条由桑隆康完成，结晶学矿物学和测试分析方法部分由赵珊茸完成，地球化学部分由李方林完成，火成岩石学、部分国际地质机构名称及缩写由肖龙完成，沉积岩石学部分由张瑞生完成。

本“术语”是在原《地球物质科学术语汇编》（桑隆康、赵珊茸、李方林、肖龙、张瑞生编，中国地质大学，2003）基础上由桑隆康和研究生文霞进行修改、补充和重新编排而成。改正了错误词条，补充了数百新词条和部分参考文献。为了方便查阅，将原“汇编”汉英地球物质科学分类术语、英汉地球物质科学术语和附录重新编排为汉英地球物质科学术语和英汉地球物质科学术语两部分，并新编了附录。研究生储玲林和汤艳参与了部分工作。

汉英地球物质科学术语部分包括 6 221 个词条，按汉语拼音字母顺序编排。英汉地球物质科学术语部分包括 6 492 个词条，按英文字母顺序排列。附录包括化学元素代号、矿物代号和国际地层表。

本“术语”标点符号中，（ ）中的内容表示解释或同义词，[ ] 中的内容表示可以省略，(?)表示该词条中文译名仅供参考。

编者在工作过程中得到中国科学院殷鸿福院士、英国岩石学家 Roger Mason 和中国地质大学龚一鸣、欧阳建平、杨坤光、鲍征宇、郑建平、曾佐勋等教授和校、院领导的大力支持和帮助，特此表示衷心感谢！

2007 年 9 月

目　录

CONTENTS

1.汉英地球物质科学术语

Chinese-English Earth Material Science Vocabularies

A

锕	Actinium(Ac)
A'KF 图	A'KF diagram
ACF 图	ACF diagram
阿尔巴尼亚	Albania
阿尔卑斯期	Alpine
阿尔卑斯型铬铁矿床	chromite deposit of Alpine-type
阿尔戈马型含铁建造	Algoma-type iron formation
阿尔及利亚	Algeria
AFM 图	AFM diagram
阿富汗	Afghanistan
阿根廷	Argentina
阿肯色州(美)	Arkansas
阿拉伯联合酋长国	The United Arab Emirates
阿拉斯加州(美)	Alaska
阿鲁巴岛	Aruba
阿曼	Oman
阿塞拜疆	Azerbaijan
阿塔苏锰矿床	Ata-Su manganese deposit
埃达克岩	adakite
埃及	Egypt
埃洛石	halloysite
埃塞俄比亚	Ethiopia
埃斯特勒双晶	Esterel twin
锿	Einsteinium(Es)
艾奥瓦州(美)	Iowa

艾伯塔省(加)	Alberta
爱达荷州(美)	Idaho
爱德华王子岛省(加)	Prince Edward Island
爱尔兰	Ireland
爱沙尼亚	Estonia
砹	Astatine(At)
安大略省(加)	Ontario
安道尔	Andorra
安多矿	auduoite
安哥拉	Angola
安圭拉岛(英)	Anguilla
安康矿	ankangite
安山岩	andesite
安提瓜和巴布达	Antigua and Barbuda
氨基酸	amino acid
氨基酸定年	amino acids age determination
氨基酸类化合物	amino acids
鞍部	saddle
鞍山式铁矿床	Anshan-type of iron deposit
鞍状矿脉	saddle reef,saddle wein
岸礁	bank reef
暗淡煤	dull coal
暗褐色	fuscous,dun
暗黑色	dingy
暗红色	wine
暗黄色	shading yellow
暗灰棕色	dull grayish brown
暗礁	reef
暗蓝色	dark blue
暗绿色	sap green
暗煤	durain
暗青绿色	gobelin blue
暗色	dark,dull,melanic,melanocratic

暗色矿物	dark-colored mineral
暗色体	melanosome
暗色岩	melanocrate
暗色岩	trap, melanocratic rock
凹槽	flute
凹凸棒石矿床	attapulgite deposit
凹凸棒石粘土矿床	attapulgite clay deposit
螯合物	chelate
奥长花岗岩	trondjemite
奥长花岗质片麻岩	trondjemitic gneiss
奥长石	oligoclase
奥地利	Austria
奥克洛现象	Oklo phenomenon
奥陶纪(系)	Ordovician Period(System), Ordovician
澳大利亚	Australia

B

BABI 值(玄武值无球粒陨石最佳初始值,代表地球形成时的 $^{87}Sr/^{86}Sr$ 值)	BABI(basaltic achondrite best initial)
八连线	octet
八面体	octahedron
八面体铁陨石	octahedrite
八面体择位能	OSPE(octahedron site preference energy)
八重晶	octet
巴巴多斯	Barbados
巴比涅补偿器	Babinet compensator
巴布亚新几内亚	Papua New Guinea
巴哈马	Bahamas
巴基斯坦	Pakistan
巴拉圭	Paraguay
巴勒斯坦	Palestine
巴林	Bahrain
巴罗带	Barrovian zones

巴拿马	Panama
巴韦诺双晶(曾用名:巴温诺双晶)	Baveno twin
巴西	Brazil
巴西双晶	Brazil twin
钯	Palladium(Pd)
靶区优选	target selection
坝	barrier
白矮星	white dwarf
白俄罗斯	Belarus
白垩纪(系)	Cretaceous Period(System)
白垩矿床	chalk deposit
白垩系	Cretaceous
白岗岩	alaskite
白磷钙石	whitlockite
白榴石	leucite
白榴岩	leucitite
白片岩	white schist
白铅矿	cerussite
白色	white
白铁矿	marcasite
白钨矿	scheelite
白云鄂博矿	baiyuneboite
白云母	muscovite
白云石	dolomite
白云石化	dolomitization
白云石化作用	dolomitization
白云岩	dolomite,dolomitite,dolostone
白云岩矿床	dolomite deposit
百幕大群岛(英)	Bermudas
百万分之一(10^{-6})	ppm(part per million)
柏林蓝	Berlin blue
斑点板岩	spotted slate
斑点构造	spotted structure

斑点消光	mottled extinction
斑点状构造	spotted structure
斑晶	phenocryst
斑块	clot
斑铜矿	bornite
斑岩	porphyry
斑岩钼矿床	porphyry molybdenum deposit
斑岩铜矿床	porphyry copper deposit
斑岩铜矿模式	porphyry copper model
斑杂构造	taxitic structure
斑状	porphyritic
斑状变晶结构	porphyroblastic texture
斑状结构	porphyritic texture
搬运[作用]	transportation
板,(棘皮动物、海参等)棘,板	sclerodermite
板块	plate
板块边界应变速率	plate boundary strain rate
板块边缘	plate margin
板块俯冲	plate subducting
板块构造	plate tectonics
板块构造成矿模式	plate-tectonic model for metallogenesis
板块碰撞	plate collision
板块运动	plate movement
板面	Pinacoid
板内岩石	within-plate rock
板内应变	intraplate strain
板劈理	slaty cleavage
板岩	slate
板褶皱	bending fold
板状构造	slaty structure
板状交错层理	tabular cross-bedding, plate cross-bedding
板状铀矿体	tabular U-ore body
半暗煤	semidull coal

半金属光泽	submetallic luster
半晶质	hemicrystalline
半径	radius
半亮煤	semibright coal
半面体	hemihedron
半磨圆的	subrounded
半泥质片岩	semi-pelitic schist
半泥质千枚岩	semi-pelitic phyllite
半泥质岩	semi-pelite
半坡面	hemidome
半熔岩	metatexite
半深海	bathyal
半深海沉积物	bathyal sediment
半深海的	bathyal
半衰期	half-life
半丝质体	semifusinite
半透明	translucent，trans-opaque
半远洋的	hemipelagic
半远洋岩	hemipelagite
半自形变晶的	hypidioblastic
半自形粒状	hypidiomorphic granular
半自形粒状结构	hypidiomorphic granular texture
伴生	associate
伴生矿	associated ore
伴生矿物	associate mineral
伴生气	associated gas
包裹体	inclosure，inclusion
包含结构	poikilitic texture
包卷层理	convolute bedding
包络面	enveloping surface
包气带水	suspended water
包体	enclosure，enclave
包头矿	baotite

孢子体	sporinite
雹痕	hail print
宝石	gemstone,jade
宝石矿产	precious stone
宝石矿物	gem mineral
宝石矿物学	gem mineralogy
宝石学	gemology
饱和[度]	saturation
饱和岩	saturated rock
饱水带	zone of saturation
保加利亚	Bulgaria
堡礁	barrier reef
鲍林法则	Pauling's rules
鲍玛序列	Bouma sequence
暴涨	inflation
爆发岩屑	tephra
爆发指数	explosive index
爆裂法	decrepitation method
爆破角砾岩型铀矿	U-ore of explosion-breccia type
爆炸性喷发	explosive eruption
北半球回归线	NHRL(nothern hemisphere regression line)
北达科他州(美)	North Dakota
北京[直立]人	Peking man
北京蓝(头青)	Peking blue
北卡罗来纳州(美)	North Carolina
北马里亚纳群岛自由联邦(美)	The Commonwealth of the Northern Mariana Islands
贝尔纳图	Bernal chart
贝壳灰岩	coquina
贝壳状断口	conchoidal fracture
贝克线	Becke line
贝克线色散	dispersion of Becke line
贝劳共和国	The Republic of Belau

贝累克补偿器(曾用名:贝瑞克补色器)	Berek compensator
贝尼奥夫带	Benioff zone
贝宁	Benin
背景值	background value
背散射电子显微镜	Backscattered Electron Microscopy (BSEM)
背散射电子像	Backscattered Electron Imaging (BEI, BSEI)
背散射扫描电子图像	Backscattered Scanning Electron Imagery (BSE)
背斜	anticline
背斜构造理论	anticline theory
背形	antiform
钡	Barium(Ba)
钡长石	celsian
钡闪叶石	barytolamprophyllite
钡铁钛石	bafertisite
钡霞石	hexacelsian
被动大陆边缘盆地	passive margins basins
被膜状集合体	filmy aggregate
本征矢量	eigenvector
崩解	disintegration
崩塌	devolution, slacktip, rockfall
比较晶体化学	comparative crystal chemistry
比较行星学	comparative planetology
比利时	Belgium
比例尺	scale
比例模拟	scale model
比重	specific gravity
比重瓶	pycnometer, Bottle Pycnometer
比重天平	Jolly Balance (JB)
比重液	Heavy Liquids
闭合度	closure

闭塞环境的	euxinic
闭形	closed form
铋	Bismuth(Bi)
铋矿床	bismuth deposit
碧绿	verdure
碧玄岩	basanite
边界品位	cutoff grade
边滩	point bar
边缘	fringe
边缘海	epeiric sea
边缘海盆[地]	marginal-sea basin
边缘蚀变	fringe alteration
变安山岩化	meta-andesitization
变斑晶	porphyroblast
变彩	labradorescence, play of color, schiller
变[质生]成矿床	metamorphogenic deposit
变辉长岩	meta-gabbro
变基性岩	metabasite
变胶体矿物	metacolloidal mineral
变胶状构造	metacolloidal structure
变角自旋核磁共振光谱	Magic-Angle Spinning Nuclear Magnetic Resonance Spectroscopy (MAS－NMR)
变晶	blast
变晶鲕	metacrystal ooid
变晶结构	blastic texture
变晶系	crystalloblastic series
变粒岩	leptynite, leptite
变嵌晶	poikiloblast
变生非晶化	metamictization
变网阳离子	network-modifier cation
变形	deformation, distortion
变形(指晶体形状)	various form
变形沉积构造	deformational sedimentary structures

变形带	deformation band
变形结构	deformation texture
变形砾石	deformation boulder
变形双晶	deformation twins
变形纹	deformation lamellae
变余斑状结构	blastoporphyritic texture
变余层理	blastobedding
变余构造	palimpsest structure
变余火成碎屑结构	blastopyroclastic texture
变余结构	palimpsest texture
变余砾状结构	blastopsephitic texture
变余糜棱岩	blastomylonite
变余砂状结构	blastopsammitic texture
变质[作用]	metamorphism
变质成矿作用	metamorphic metallization
变质带	metamorphic belt,metamorphic zone
变质反应	metamorphic reaction
变质分泌成矿作用	metallization by metamorphic secretion
变质分异	metamorphic differentiation
变质构造	metamorphic structure
变质级	metamorphic grade
变质阶段(幕)	metamorphic episode
变质结构	metamorphic texture
变质结晶[作用]	metamorphic crystallization
变质矿床	metamorphic mineral deposit
变质期	metamorphic epoch
变质热梯度	metamorphic thermal gradient
变质事件	metamorphic event
变质体制	metamorphic regime
变质相	metamorphic facies
变质相系	metamorphic facies series
变质相组	metamorphic facies group
变质旋回	metamorphic cycle

变质亚相	metamorphic subfacies
变质岩	metamorphic rock
变质作用	metamorphism
变质作用 *P/T* 比类型	*P/T* ratio type of metamorphism
变质作用类型	type of metamorphism
变种(矿物分类单位)	variety
辫状河流	braided stream
标本	specimen
标型矿物	typomorphic mineral
标型特征	typomorphic characteristic
标型性	typomorphism
标型元素	typochemical element
标志层	key bed
表[皮]土	regolith
表鲕	superficial ooid
表观年龄	apparent age
表壳岩	supracrustal rock
表面扩展 X 射线吸收精细结构光谱	Surface-Extended X-ray Absorption Fine-Structure Spectroscopy (SEXAFS)
表生矿物	hypergene mineral, supergene mineral
表生碎屑	epiclast
表生异常	hypergene anomaly
表生作用	supergenesis, hypergenesis
宾夕法尼亚亚系	Pennsylvanian
宾夕法尼亚州(美)	Pennsylvania
滨海	shore
滨外沙坝	barrier bar
滨外滩	barrier beach
冰雹	hailstone
冰长石	adularia(adular)
冰长石化	adularization
冰长石晕彩	adularescence
冰川	glacier

冰川擦痕	glacial stria
冰川地貌	glacial landform
冰川湖	glacier lake
冰川漂砾	glacial erratic boulder
冰川作用	glaciation
冰岛	Iceland
冰斗	glacial cirque
冰筏	ice raft
冰盖	ice cap
冰后期	post-glacial period
冰晶石	cryolite
冰溜面	glacial pavement
冰劈作用	frost splitting
冰期	glacial stage, ice age
冰碛	moraine, till, glacial drift
冰碛地貌	till landform
冰碛阶地	moraine terrace
冰碛扇	moraine fan
冰碛物	till
冰染蓝	ice blue
冰蚀谷	glacial valley
冰蚀作用	glacial erosion
冰水沉积平原	outwash plain
冰水堆积	glacial fluvial deposit
冰水扇	outwash fan
冰洲石	icespar
波长分散 X 射线光谱	Wavelength Dispersive X-ray Spectroscopy (WDXS)
波长色散光谱	Wavelength Dispersive Spectroscopy (WDS)
波动说	undation theory
波多黎各岛(美)	Puerto Rico
波峰	ripple crest

波痕	ripple mark
波脊线	wave crest line
波兰	Poland
波浪折射	wave refraction
波切台	marine-cut platform
波斯尼亚和塞尔维亚	Bosnia and Hercegovina
波斯特马斯堡锰矿床	Postmasburg manganese deposit
波状交错层理	wavy cross-bedding
波状消光	undulating extinction, undośe extinction
玻恩-梅尔理论	Born-Mayer theory
玻基斑状结构	vitroporphyritic texture
玻晶交织结构	hyalopilitic texture
玻璃	glass
玻璃光泽	glassy luster, vitreous luster
玻璃糜棱岩	hyalomylonite
玻璃原料	glass raw materials
玻璃原料矿产	raw material for glass
玻璃陨石	tektite
玻璃质	amorphous substance, glassy, vitreous
玻璃质碎屑结构	vitriclastic texture
玻利维亚	Bolivia
玻屑	glass shard , vitric pyroclast
玻陨石	tectite
玻质碎屑岩	hyaloclastite
剥离丘	exfoliation dome
剥蚀	disintegration
剥蚀[作用]	denudation
剥蚀平原	plain of denudation
伯利兹	Belize
铂	Platinum(Pt)
铂系(族)元素	platinum group element(PGE)
铂族元素矿床	ore deposit of platinum group
博茨瓦纳	Botswana

勃吉尔德连晶(斜长石出溶形成)	Bϕggild intergrowth
薄板状	platy
薄片	thin section
薄片鉴定法	thin section method
簸选	winnowing
补色器	compensator
新型	neotype
捕虏体	xenolith
不饱和岩	unsaturated rock
不变点	invariant point
不变平衡	invariant equilibrium
不变子群	invariant subgroup
不丹	Bhutan
不等粒斑状结构	inequigranular porphyritic texture
不等粒结构	heterogranular texture
不等粒状	inequigranular
不对称波痕	asymmetrical ripple
不对称褶皱	asymmetrical fold
不共轭	non-conjugacy
不混溶性	immiscibility
不混溶岩浆	immiscible magma
不可逆的(单向的)转变	monotropic transformation
不连续	discontinuity
不连续反应	discontinuous reaction
不连续相变	discontinuous transformation
不列颠哥伦比亚省(加)	British Columbia
不平衡结构	disequilibrium texture
不透明	[translucent] opaque
不透明	opaque
不完全解理	poor cleavage ,imperfect cleavage,less good cleavage
不完全类质同象	incomplete isomorphous series
不稳[定状]态	unstable state

不稳定性	unstability
不相称结构	incommensurate structure
不相容元素	incompatible elements
不谐调褶皱	disharmonic fold
不一致年龄	discordia age
不整合	unconformity
不整合脉型铀矿	unconformity-vein type uranium deposit
不整合侵入体	discordant intrusion
不整合圈闭	unconformity trap
布伽重力异常	Bouguer gravity anomaly
布格重力异常	Bouguer gravity anomaly
布基纳法索	Burkina Faso
布拉格定律	Bragg's law
布拉格方程	Bragg equation
布拉维法则	law of Bravais
布拉维晶格	Bravais lattice
布拉维指数	Bravais indices
布隆迪	Burundi
布容极性时	Brunhes polarity chron
布儒斯特定律(曾用名:布鲁斯特定律)	Brewster's law
布什维尔德铬矿床	Bushveld chromium deposit
钚	Plutonium(Pu)
部分平衡	partial equilibrium
部分熔融	partial melting

C

参数(量)	parameter
参数井	key well, stratigraphic well
残积和机械富集作用	residual and mechanical concentration
残积矿床	residual deposit
残积砂矿	eluvial placer
残积物	eluvium, residual sediments

残留磁性测量	Remnant Magnetism Measurements (RMM)
残留矿床	residual deposit
残缕结构	helicitic texture
残余构造	relict structure
残植(殖)煤	liptobiolite
糙面	shagreen surface
槽	trough, flute
槽沟	groove
槽痕	flute marks
槽台成矿说	geosyncline-platform theory for metallogenesis
槽状交错层理	trough cross-bedding
草图	sketch
CDT 标准(硫同位素标准,美国亚利桑那铁陨石中的陨硫铁的硫同位素比值)	CDT standard(Canyon Diablo meteorite troilite standard)
侧分泌说	lateral secretion theory
参差状断口	uneven fracture
侧碛堤	lateral moraine bar
测角仪	goniometer
测线采样	profile sampling
测重湿化学分析法	Gravimetric-Wet Chemical Analysis (GWCA)
层	bed
层(条带)状混合岩	stromatite
层化	stratification
层火山	stratovolcano
层间构造	internal bedding structure
层间水	interlayer water
层孔虫类	stromatoporoid
层控矿床	strata-bound mineral deposit
层理	stratification, bedding

层裂	exfoliation
层流	laminar flows
层面	bedding plane
层侵系	Rhyacian
层位	horizon
层序加积的	aggradational
层状剥落	bed separation
层状构造	layered structure, stromatic structure
层状硅酸盐	phyllosilicate(sheet silicate)
层状火山	stratovolcano
层状孔洞构造	stromatactis
层状矿床	stratiform [mineral] deposit
层状铜矿床	stratiform copper deposit
C-格子	C-lattice, base-centered lattice
插入	intercalation
茶[褐]色	tawny, dark brown, tan
差热分析	Differential Thermal Analysis(DAT)
差式扫描量热分析	Differential Scanning Calorimeter (DSC)
差异风化	differential weathering
拆沉	delamination
柴达木石	chaidamuite
产层有效厚度	net pay thickness
产油层	reservoir
产油区	oil-producing region
产铀岩体	uranium productive massif
产状	attitude
长板状	lath shaped
长石	feldspar
长石矿床	feldspar deposit
长石砂岩	arkose, feldspar sandstone
长石石英砂岩	feldspathic quartz sandstone
长石岩屑砂岩	feldspar lithic sandstone
长焰煤	long-flame coal

长英角岩	felsic hornfels
长英麻粒岩	leptite
长英质	felsic, quartz-feldspathic
长英质岩石	felsic rock
肠状混合岩	ptygmatic migmatite
肠状岩	ptygmatite
常光	ordinary ray
常数	coefficient
常温层	constant zone of subsurface temperature
场发射扫描电子显微镜	Field-emission Scanning Electron Microscopy (FESEM)
场致发光	electroluminescence
超变质[作用]	ultrametamorphism
超变质岩	ultrametamorphic rock
超地幔柱	super plume
超高压	ultra-high pressure(UHP)
超高压变质岩	ultra-high-pressure metamorphic rock
超基性	ultrabasic
超基性片麻岩	ultrabasic gneiss
超基性岩	ultrabasic rock
超结构	super structure
超晶胞	super-cell
超晶格	superlattice
超临界含矿流体	supracritical ore fluid
超临界状态	super-critical state
超镁铁岩	ultramafic rock
超镁铁质	ultramafic
超镁铁质岩	ultramafic rock
超糜棱岩	ultramylonite
超密说	superdense hypothesis
超声波干扰(干涉)测量	Ultrasonic Interferometry (UI)
超酸性岩	ultraacidic rock
超碎裂岩	Ultracataclasite

超显微的	ultramicroscopic
超显微条纹长石(电子显微镜下)	cryptoperthite
超新星爆发	supernova explosion
超盐度	hypersaline
超营养作用	eutrophication
朝鲜民主主义人民共和国	The Democratic People's Republic of Korea
潮差	tidal range
潮间带	intertidal zone
潮坪(滩)	tidal flat
潮上滩	sabkha
潮汐	tide
潮下带	subtidal zone
尘暴	dust storm, dust bowl, dust devil
沉淀[作用]	precipitation
沉淀分带	precipitation zoning
沉积[作用]	depositional process, deposition, precipitation, sedimentation
沉积变质矿床	sedimento-metamorphic deposit
沉积变质磷矿床	sedimento-metamorphic phosphate deposit
沉积变质锰矿床	sedimento-metamorphic manganese deposit
沉积成岩作用	diagenesis
沉积环境	sedimentary environment
沉积建造	depositional setting
沉积矿床	sedimentary deposit
沉积喷流矿床	sedimentary exhalative deposit
沉积盆地	sedimentary basins
沉积体系	sedimentary systems
沉积铁矿床	sedimentary iron deposit
沉积同期的(与沉积同时的)	syndepositional
沉积物	deposition

沉积相	sedimentary facies
沉积学	sedimentology
沉积岩	sedimentary rock
沉积作用	sedimentation
沉降	subsidence
沉降速度	settling velocity
辰砂	cinnabar
陈化	ageing
衬度	contrast
成冰系	Cryogenian
成层	bedding
成分层	compositional layering
成分成熟度	compositional maturity
成分-共生图解	composition-paragenesis diagram
成核	nucleation
成矿带	metallogenic belt
成矿建造	metallogenic formation
成矿阶段	stage of mineralization
成矿壳层	ore-forming level
成矿控制因素	metallogenic factor
成矿模式	metallogenic model
成矿期	metallogenic epoch,epoch of mineralization
成矿区	metallogenic region
成矿省	metallogenic province
成矿时代	metallogenic(metallogenetic) epoch
成矿系(序)列	minerogenetic series,metallogenic series
成矿学	metallogeny
成矿预测	metallogenic prediction
成矿域	metallogenic domain,metallogenic megaprovince
成矿元素	ore-forming element
成矿专属性	metallogenic specialization

成矿作用	metallogenesis, minerogenesis, ore-forming process
成煤物质	coal-forming material
成熟	mature, maturation
成熟度	maturity
成铁系	Siderian
成土因素	factor of soil formation
成网阳离子	network-former cation
成岩[作用]	diagenesis
成岩[作用]的	diagenetic
成岩格子	petrogenetic grid
成岩阶段	diagenetic stages
成因矿物学	genetic mineralogy
承压水	artesian water
承载能力	carrying capacity
程序控制 X 射线层析	Computed X-ray Tomography(CXT)
澄江动物群	Chengjiang fauna
橙红	orange red, salmon pink
橙色	orange
橙玄玻璃	palagonite
橙玄岩	palagonite
橙棕色	orange brown
持续能力	sustainable capacity
尺度模拟	scale model
赤道几内亚	Equatorial Guinea
赤褐色	henna, mahogany, rufous, russet, russety, bronze
赤红	crimson
赤路矿	chiluite
赤铁矿	hematite
赤铁矿化	hematitization
赤铜矿	cuprite
炽热火山云	nuée ardente, glowing cloud

充填[作用]	cavity filling
充填矿床	mineral deposit by cavity filling
充填物	matrix
冲槽	flute
冲击变质[作用]	impact metamorphism
冲击变质岩	impact metamorphic rock
冲击玻璃	impactite glass
冲击角砾岩	suevite
冲击熔岩	impact lava
冲击岩	impactite
冲积层	alluvion
冲积平原	alluvial plain
冲积砂矿	alluvial placer
冲积扇	alluvial fan
冲刷构造	scour structure
冲刷痕	scour marks
冲洗交错层理	swash cross-bedding
虫迹	tracks
重结晶[作用]	recrystallization,recrystallize
虫孔	burrows
重新分配(重新分布)	redistribute
重新活动	remobilization
重折率	birefringence
畴结构	domain structure
臭葱石	scorodite
臭氧	ozone
臭氧层	ozonosphere
臭氧耗竭	ozone depletion
出溶	exsolution,exsolved
出溶结构	exsolution texture
出溶条纹	exsolution lamella
初波(P 波)	primary wave(P wave)
初级生产	primary production

初糜棱岩	protomylonite
初始比值	initial ratio
初始铅	initial lead
初始熔融	initial melt
初碎裂岩	protocataclasite
雏晶	crystallite
储层性质	reservoir property
储存性资源	stock resource
储集层	reservoir
储量	reserve
储油地质学	reservoir geology
储油构造	oil-bearing structure
储油气层	reservoir formation
穿插双晶	penetrate twin
穿时的	diachronous
穿时递增变质[作用]	diachronous-progressive metamorphism
传导	conduction
传导热流模拟	conductive heat flow modeling
传导性	conductivity
传压介质	pressure transmitting medium
吹蚀	winnowing
吹蚀作用	deflation
CHUR 值(球粒陨石型均一岩浆房的地幔库)	CHUR(chondritic uniform reservoir)
垂向序列	vertical succession
垂直分带	vertical zoning
垂直分带性	altitudinal zonality
垂直双晶	normal twin
垂直运动	vertical movement
纯橄榄岩	dunite
瓷板	porcelain streak plate
磁暴	magnetic storm
磁赤铁矿	maghemite

磁化率	magnetic susceptibility
磁黄铁矿	pyrrhotite
磁力显微镜	Magnetic Force Microscopy (MFM)
磁偏角	magnetic declination
磁倾角	magnetic inclination
磁铁矿	magnetite
磁铁石英岩	magnetite quartzite
磁性	magnetism
雌黄	orpiment
次变边结构	kelyphitic texture
次级结构单元	secondary building units
次级生产(加工工业)	secondary production
次棱角状的	subangular
次生包体	secondary inclusion
次生富集带	zone of secondary enrichment, secondary enriched zone
次生富集作用	secondary enrichment
次生环境	secondary environment
次生节理	secondary joint
次生孔隙度	secondary porosity
次生矿物	secondary mineral
次生石英岩	secondary quartzite
次生双晶	secondary twin
次生异常	secondary anomaly
次透辉石	salite
次元素	minor element
次圆状的	subrounded
刺	sclerodermite
粗安岩	trachyandesite
粗糙的	coarse
粗糙断口	uneven fracture
粗砾岩	cobble conglomerate
粗粒的	coarse-grained

粗粒结构	coarse grained texture
粗粒体(烟煤和褐煤显微组分)	macrinite
粗粒杂砂岩	coarse wacke
粗面结构	trachytic texture
粗面岩	trachyte
粗铅法	rough lead method
粗砂	grit
粗砂岩	coarse sandstone, grit
粗尾粒序	coarse-tail grading
粗玄岩	trachybasalt
脆性	brittleness
脆性变形	brittle deformation
脆性断裂	brittle fracture
脆性行为	brittle behavior
淬火	quench
淬火熔体	quench melt
翠绿	emerald

D

大爆炸	big bang
大爆炸宇宙学	big-bang cosmology
大地构造学	tectonics
大地水准面	geoid [surface]
大骨节病	Kaschin-Beck disease
大红	bright red, scarlet, crimson
大矿囊	bonanza
大类(矿物分类单位)	class
大离子亲石元素	large ionic lithophile element (LILE)
大理岩	marble
大陆	continent
大陆[性]冰川	continental glacier
大陆裂谷系	continental rift system
大陆漂移	continental drift

大陆漂移说	continental drift theory
大陆热流	continental heat flow, heat flow through the continent
大气降水	meteoric waters
大气圈	atmosphere
大气氩	atmospheric argon
大青山矿	daqingshanite
大洋中脊	mid-oceanic ridge, mid-ocean ridge
大洋中脊玄武岩	mid-oceanic-ridge-basalt (MORB)
大隅石	osumilite
代	Era
带褐色的	brownish
带黄色的	yellowish
带绿色的	greenish
带青色的	blueish
带隙	band gap
带状沉积	banded sediment
带状构造	layered structure, stromatic structure
丹巴矿	danbaite
丹麦	Denmark
单变平衡	univariant equilibrium
单变线	univariant curve
单成分砾岩	oligomictic conglomerate, monogenetic conglomerate
单成分粘土岩	monogenetic claystone
单分子物体	monomer
单键型晶格	homodesmic lattice
单阶段体系	single stage system
单晶	monocrystalline
单晶 X 射线衍射	Single-Crystal X-ray Diffraction(SCXRD)
单晶精细 X 射线分析	Single-Crystal Refinement X-ray Analysis(SREF-XRA)
单聚物	monomer

单颗粒锆石离子探针质谱法	SHRIMP(The zircons were analyzed by a specially built high resolution ion microprobe at the Australian National University nicknamed 'SHRIMP')
单矿岩	monomineralic rock
单链状硅酸盐	single chain silicate
单面	pedion
单偏光	plane-polarized light(ppl)
单圈测角仪	one circle goniometer
单水型的	monohydrate
单位晶胞"分子"数(或简称 Z 数)	Zahl[德](或简称 Z)
单向的(不可逆的)转变	monotropic transformation
单斜[层]	monocline,homocline
单斜辉石	clinopyroxene
单斜辉石岩	clinopyroxenite
单斜晶系	monoclinic system
单斜钠长石	monalbite
单形	single form
单形符号	single form symbol
单质	element
单锥	pyramid
淡橙色	light orange
淡赤黄色	ginger
淡粉红色	rose pink
淡褐色	pale brown,light brown,maple,hazel
淡红色	rosiness,damask
淡黄	yellowy, canary, jasmine, straw yellow, wheat,flaxen,primrose yellow
淡黄褐色的	fawn-coloured
淡黄绿色	pistachio
淡灰褐色	ficelle
淡蓝	pale blue,light blue,baby blue

淡绿色	pale green, light green, pea green, absinthe-green, virescent, viridescent
淡青色	nattierblue
淡色	pale, tinge, light, light-colored, leucocratic
淡银灰色	ocean-grey
淡紫兰色	lavender blue
淡紫色	heliotrope, lavender, lilac, orchid
淡棕色	pale brown, light brown, maple
蛋白光	opalescence
蛋白石	opal
蛋白石化	opalization
蛋白石矿床	opal deposit
蛋白质	protein
氮	Nitrogen(N)
氮循环	nitrogen cycle
导带	conduction band
导电性	conductivity, electrical conductivity, conductivity
岛弧	island arc
岛弧系	island-arc system
岛弧岩石	island arc rock
岛屿结构	island texture
岛状硅酸盐	nesosilicate
倒石堆	talus
倒易格子	reciprocal lattice
倒置(转)	inversion
倒转变质带	inverted metamorphic zones
倒转层序	reversed succession, reversed order
delta 值[绝对变差 δ,反映同位素组成与标准相比重同位素的富集或亏损程度(‰)]	delta value
倒转褶皱	overturned fold
道马矿	daomanite

得克萨斯州(美)	Texas
锝	Technetium(Tc)
德拜-谢勒法	Debye-Scherrer method
德国	Germany
等变线	isograd
等反射率线	isoreflectance line
等分线	bisectrix
等化学变质[作用]	isochemical metamorphism
等化学变质岩	isochemical metamorphic rock
等化学的	isochemical
等价类质同象	isovalent isomorphism
等结构	isostructure
等离子体	plasma
等粒结构	even-grained, homogranular texture
等粒状	equigranular
等煤级线	isorank
等倾斜(角)线	dip isogon
等倾斜线作图法	isogon construction
等热	isotherm
等时线截距	intercept of isochron
等时线拟合参数	mean square of weighted deviates(MSWD)
等时线年龄	isochron age
等时线斜率	slope of isochron
等温线	isotherm
等物理变质岩	isophysical metamorphic rock
等效点系	equipoints
等斜(同斜)褶皱	isoclinal fold
等型	isotype
等压线	isobar
等震线	isoseismal line
等轴的	equidimensional
等轴晶系	isometric system, cubic system
低级变质岩	low-grade metamorphic rock

低角度逆冲断层	low angle reverse thrust
低能电子衍射	Low-Energy Electron Diffraction (LEED)
低速带	low velocity zone
低突起	low relief
低温地球化学	low temperature geochemistry
低温热液矿床	epithermal deposit
低温水热矿床	epithermal deposit
低压相系	low-pressure facies series
镝	Dysprosium(Dy)
迪尔石	Deerite
迪开石	Dickite
底[轴]面	basal pinacoid
底辟构造	diapir structure
底痕	sole marks
底浸	underplating
底面	subface
底面双晶	acline twins
底栖的	benthonic
底碛平原	ground moraine plain
底碛丘陵	ground moraine mound
底心格子	base-centered lattice,C-lattice
地表的	terrestrial
地表生态系统	terrestrial ecosystem
地表形态学	morphology of the earth surface
地槽	geosyncline
地槽型成矿建造	metallogenic formation of geosyncline type
地层	stratum,bed
地层地球化学	stratigraphic geochemistry
地层对比	stratigraphic correlation
地层剖面	stratigraphic section
地层圈闭	stratigraphic trap
地层学	stratigraphy

地层柱	stratigraphic column
地磁负向场	reversed field
地磁极	geomagnetic pole
地磁极性倒转	geomagnetic reversal
地磁异常	magnetic anomaly
地磁正向场	normal field
地方病	endemic disease
地方时	local time
地核	[earth] core
地核差异旋转	differential rotation of earth core
地极	earth pole
地壳	[earth] crust
地壳低速带	crustal low-velocity zone
地壳元素丰度	clark value
地壳运动	crustal movement
地蜡	ozocerite
地垒	horst
地理学	geography
地沥青	land asphalt
地裂缝	ground fracture
地幔	mantle
地幔部分熔融	partial melting of mantle
地幔地球化学	mantle geochemistry
地幔对流	mantle convection
地幔岩	pyrolite
地幔柱	[mantle] plume
地貌	landform, physiognomy
地貌内营力	internal geomorphic agency
地貌外营力	external geomorphic agency
地貌学	geomorphology
地貌演化	geomorphic evolution
地貌营力	geomorphic agency
地面沉降	ground setting

地堑	graben
地球	earth
地球表层科学	earth surface sciencc
地球磁性	geomagnetism
地球动力学	geodynamics
地球化学	geochemistry
地球化学背景	geochemical background
地球化学参考模型	geochemical earth reference model(GERM)
地球化学分散	geochemical dispersion
地球化学封闭体系	geochemical closed system
地球化学环境	geochemical environment
地球化学活动性	geochemical mobility
地球化学建模	geochemical modeling
地球化学开放体系	geochemical open system
地球化学勘查	goechemical exporation
地球化学模型	geochemical model
地球化学普查	geochemical reconnaissance
地球化学省	geochemical province
地球化学输运	geochemical transport
地球化学探矿	geochemical prospecting
地球化学体系	geochemical system
地球化学填图	geochemical mapping
地球化学图	geochemical chart
地球化学污染调查	geochemical survey of pollution
地球化学相	geochemical facies
地球化学详查	geochemical detailed survey
地球化学旋回	geochemical cycle
地球化学演化	geochemical evolution
地球化学异常	geochemical anomaly
地球化学障	geochemical barrier
地球化学自组织	geochemical self-organization
地球科学	earth science,geoscience
地球内部化学	interior chemistry of earth

地球物理学方法	geophysical technique
地球物质科学	earth material science
地球系统	earth system
地球系统科学	earth system science
地球震荡	earth oscillation
地球子午圈	terrestrial meridian
地热	geotherm
地热地球化学	geothermal geochemistry
地热梯度	geothermal gradient
地热系统	geothermal system
地台	platform
地台型成矿建造	metallogenic formation of platform type
地体	terrain
地洼型成矿建造	metallogenic formation of diwa (geodepression)type
地外变质[作用]	extraterrestrial metamorphism
地外文明	extraterrestrial civilization
地温	geotemperature
地下构造	sub-surface structure
地下热[卤]水型矿床	hot geothermal brine type deposit
地下水	ground water,groundwater,underground water,phreatic water
地下水循环成矿模式	groundwater-circulation metallogenic model
地形侵蚀回春	rejuvenated landform
地形同位素效应	topographic isotopic fractionation effect
地形雨	orographical rain
地学	geoscience,earth science
地衣	lichen
地域分异	regional differentiation
地缘政治造成的资源短缺	resource scarcity due to geopolitics
地震	earthquake,seism
地震波	seismic wave
地震带	seismic zone ,Benioff zone

地震的	seismological
地震烈度图	earthquake intensity map
地震区划图	seismic zoning map
地震效应	earthquake effect
地震序列	seismic sequence
地震预报	earthquake prediction
地震预报要素	elements of earthquake prediction
地震震级	magnitude of earthquake
地震作用	seismism
地植物方法	geobotanical method
地志学	topology
地志学的	topological
地质[学]	geology
地质边界	geological boundary
地质构造	geologic structure
地质露头	geological outcrop
地质年代表	geochronologic scale, geologic time scale
地质年代单位	geochronologic unit
地质年代学	geochronology
地质事件	geological event
地质体	geologic body
地质图	geological map
地质图解析	map interpretation
地质温度测定法	geothermometry
地质温度计	geothermometer, geologic thermometer
地质温压计	geothermobarometer
地质循环	geological cycle
地质压力测定法	geobarometry
地质压力计	geobarometer, geologic barometer
地质作用(过程)	geologic process
地轴	earth axis
递(渐)进变形	progressive deformation
递变层理	graded bedding

递进成矿说	progressive metallogenesis
递增变质	progressive metamorphism
第三纪(系)	Tertiary Period(System)
第四纪(系)	Quaternary Period(System), Quaternary
第四纪冰期	Quaternary ice age
碲	Tellurium(Te)
碲矿床	tellurium deposit
典型结构	typic structure
点群	point group
点沙坝	point bar
碘	Iodine(I)
碘矿床	iodine deposit
电磁场	earth magnetic field, geomagnetic field
电磁场倒转	reversal of geomagnetic field
电磁性	electromagnetism
电负性	electro negativity, electronegativity
电感偶合等离子分析	Inductively Coupled Plasma Analysis(ICP)
电感偶合等离子质谱仪	Inductively Coupled Plasma-Mass Spectroscopy (ICP-MS)
电离电势(位)	ionisation potential
电离度	ionicity
电气石	tourmaline
电气石化	tourmalinization
电容率	dielectric constant
电致伸缩	electrostriction, converse piezoelectricity
电轴	piezoelectric axis
电子构型	electron configuration
电子能损光谱	Electron Energy-Loss Spectroscopy (EELS)
电子亲和性	electron affinity
电子顺磁共振光谱	Electron Paramagnetic Resonance Spectrometry (EPR, EPRS)

电子探针分析	Electro Microprobe Analysis (EMPA)
电子探针穆斯保尔光谱	Electron Microprobe Analysis-Mossbauer Spectroscopy (EMPA/Mossbauer)
电子衍射	Electron Diffraction (ED)
电子自旋	Electron Spin Resonance (ESR)
电子自旋共振光谱	electron spin resonance spectroscopy
玷污	contamination
靛[青]	indigo
靛酚蓝	indophenol blue
靛蓝	indigo blue,thumb blue
钓鱼岛石	diaoyudaoite
叠层构造	stromatolitic structure
叠层生物孔层构造	stromatactis
叠层石	stromatolite
叠层石灰岩	stromatolitic limestone
叠积[作用]	aggradation
叠加异常	superimposed anomaly
叠加褶皱	superposed fold
叠生矿床	telescoped deposit
叠生作用	telescoping
叠套矿床	telescoped deposit
叠套作用	telescoping
叠瓦构造	imbricate structure,imbricated structure
叠瓦扇	imbricate fan
叠瓦式断层	imbricate fault
顶	apex
顶峰变质条件	peak metamorphic condition
顶级群落	climax
顶面	superface
定量模拟	quantitative model
定向附生	epitaxy
定向附生的	epitaxic,epitaxial

定向构造	directional structure
定向切片	oriented section
定向压力	directed pressure
定形	constant form
铥	Thulium(Tm)
东帝汶	East Timor
东萨摩亚(美)	Eastern Samoa
东太平洋洋隆成矿作用	submarine metallization at East Pacific Rise
氡	Radon(Rn)
氡法测量	radon measurement
动力变质[作用]	dynamic metamorphism
动力变质岩	dynamometamorphic rock
动力学	dynamics
动热变质岩	dynamothermal metamorphic rock
动态平衡	homeostasis
动态重结晶	dynamic recrystallization
冻蓝闪石	barroisite
冻融作用	cryopedological action
冻土地貌学	geocryology
冻土学	cryopedology
洞穴	cave
豆荚状	podiform
豆石	pisolite
豆状构造	pisolitic structure
豆状集合体	pisolitic aggregate
毒砂	arsenopyrite
独居石	monazite
端面	pedion
短波红外光谱	Short-Wave Infrared Spectroscopy (SWIS)
短期预报	short-range forecast
短轴褶皱	brachyfold
短柱状	stumpy

段	member
断层	fault
断层(裂)带	fault zone
断层变质岩	dislocation metamorphic rock
断层地震	fault earthquake
断层几何学	fault geometry
断层角砾岩	dislocation breccia,fault breccia
断层面	fault surface
断层泥	fault clay,fault gouge
断层圈闭	fault trap
断层线	fault trace
断层岩	fault rock
断距	separation
断口	fracture
断裂	fracture
断裂带	fracture zones
断裂地貌	rift geomorphology
断裂阵列	fault array
断盘	fault wall
堆积[作用]	accumulation
堆积阶地	constructional terrace
堆积平原	plain of accumulation
堆晶岩	cumulate
对策	strategy
对称	symmetry
对称变换矩阵	symmetric operation matrix
对称波痕	symmetrical ripples
对称操作	symmetry operation
对称带状构造	symmetrical banded structure
对称面	symmetry plane
对称消光	symmetric extinction
对称型	class of symmetry
对称要素	symmetry element

对称褶皱	symmetrical fold
对称中心	center of symmetry
对称轴	symmetry axis
对流	convection
对流层	troposphere
对流循环	convective circulation
对流作用	convection
盾火山	shield volcano
盾形火山	shield volcano
多边形拼图结构	polygonal mosaic
多菲内双晶(曾用名:道芬双晶)	Dauphine twin
多哥	Togo
多硅白云母	phengite
多键型晶格	heterodesmic lattice
多阶段体系	multistage system
多金属结核	polymetallic nodule
多金属矿床	polymetallic deposit
多晶畴的	polysomatic
多晶畴现象	polysome
多晶的	polycrystalline
多晶体的	polysomatic
多晶体现象	polysome
多孔的	vuggy
多孔性	porosity
多孔状[构造]的	vesicular
多铝红柱石	mullite
多米尼加共和国	The Dominican Republic
多米尼加联邦	The Commonwealth of Dominica
多面体	polyhedron
多普勒效应	Doppler effect
多期变质[作用]	polymetamorphism
多色性	pleochroism
多色性公式	pleochroic formula, pleochroic scheme

多水氯硼钙石	hydrochlorborite
多糖	polysaccharide
多相变质岩	plurifacial metamorphic rock
多型	polytype
多型转变	polymorphic transition
多旋回成矿作用	polycyclic metallogeny
多因复成矿床	polygenetic and compound deposit
惰性气体	noble gas, inert gas, rare gas, atmophile element
惰性铀	immobile uranium
惰性组分	inert components
惰质组	inertinite
DUPAL 异常(Hart (1984)注意到具有$^{207}Pb/^{204}Pb$和$^{208}Pb/^{204}Pb$对给定的$^{206}Pb/^{204}Pb$值的高比值的大洋玄武岩主要来自集中于约 30°S 的带。Hart 将这一特征命名为 DUPAL 异常(以 Dupré 和 Allègre 命名)。他将该 DUPAL 同位素识别标志定义为较高的 ΔSr(ΔSr=[$^{87}Sr/^{86}Sr$－0.703 0]×104)以及高 Δ8/4 和 Δ7/4)	DUPAL(Hart (1984) noticed that oceanic basalts with high $^{207}Pb/^{204}Pb$ and $^{208}Pb/^{204}Pb$ ratios for a given value of $^{206}Pb/^{204}Pb$ come mainly from a belt centered at about 30° S. Hart named this feature the DUPAL anomaly (after Dupré and Allègre). He defined the DUPAL isotopic signature as having higher ΔSr(ΔSr = [$^{87}Sr/^{86}Sr$ － 0.703 0] × 104) and high Δ8/4 and Δ7/4)

E

俄亥俄州(美)	Ohio
俄克拉何马州(美)	Oklahoma
俄勒冈州(美)	Oregon
俄罗斯	Russia
峨眉矿	omeiite
锇	Osmium(Os)
额尔齐斯石	ertixiite
厄尔尼诺	El Nino
厄瓜多尔	Ecuador
厄立特里亚	Eritrea

恩福拉双晶	Emfola twin
鲕粒灰岩	oolitic limestone
鲕粒亮晶灰岩	oosparite
鲕粒泥晶灰岩	oomicrite
鲕粒岩	oolite
鲕粒岩的	oolitic
鲕绿泥石	chamosite
鲕状的	oolitic
鲕状构造	oolitic structure
鲕状集合体	oolitic aggregate
铒	Erbium(Er)
二[向]色性	dichroism
二长花岗岩	monzonitic granite
二长结构	monzonitic texture
二长片麻岩	two-feldspar gneiss
二长岩	monzonite
二次电子像	Secondary Electron Imaging (SEI)
二次高质量共振离子探针	Secondary High-Mass Resolution Ion Microprobe (SHRIMP)
二次离子质谱仪	Secondary-Ion Mass-Spectrometry (SIMS)
二叠纪(系)	Permian Period(System), Permian
二辉橄榄岩	lherzolite
二辉岩	websterite
二级相变	second order transition
二阶段体系	two stage system
二连石	Erlianite
二面角	dihedral angle
二十面体	icosahedron
二维核	two-dimensional nucleus
二维图形	stereogram
二元共晶	binary eutectic
二重成因矿床	diplogenetic deposit

二轴晶	biaxial crystal

F

发光性	luminescence
发光云	glowing cloud
发射光谱	Emission Spectroscopy (ES)
发生	generation
发夕线石	fibrolite
法国	France
法罗群岛(丹)	Faeroe Islands
法线	normals to plane
法向应力	normal stress
法属波利尼西亚	French Polynesia
法属圭亚那	French Guiana
翻转褶皱	facing dawn fold
钒	Vanadium(V)
钒钡铜矿	vesignieite
钒钙铀矿	tyuyamunite
钒酸盐	vanadate
钒钛磁铁矿矿床	vanadic titanomagnetite deposit
钒铜矿	vanadinite
钒云母	roscoelite
反磁化	reverse magnetization
反磁性	diamagnetism
反气旋	anticyclone
反射	reflectance, reflection
反射计	reflectometer
反射率	reflectance, reflectivity
反射率分析法	reflectance analysis
反射像	Reflectance Imagine (RI)
反射颜色指数	reflection color index
反射折射仪	Reflectance Refractometry (RR)
反条纹长石	antiperthite

反铁磁性	antiferromagnetism
反相畴	antiphase domains
反向环带	reverse zoning
反向期	reversed epoch
反压电性	converse piezoelectricity,electrostriction
反应边	reaction rim
反应边结构	reaction rim texture
反应等变线	reaction isograd
反应结构	reaction texture
反应力矿物	antistress mineral
反应速率	reaction rate
反应系列	reaction series
反映双面	reflection dihedron
泛大陆	Pangea
泛大洋	Panthalassa
泛滥平原	flood plain
范德华键	Van dee Waals' bond
梵蒂冈	Vatican City State
方沸石	analcime,analcidite,analcite
方格纸	squared paper
方解石	calcite
方解石化	calcitization
方镁石	periclase
方钠石	sodalite
方硼石	boracite
方铅矿	galena
方石英	cristobalite
方铁矿	wustite
方位	azimuth
方位角	bearing,orientation angle,azimuth
方陨铁	hexahedrite
方柱石	scapolite
方柱石化	scapolitization

钫	Francium(Fr)
放大镜	hand lens
放射虫	radiolarian
放射虫岩	radiolarite
放射鲕	radial ooid
放射性测量	radioactive measurement
放射性示踪分析	radiotracer analysis
放射性衰变	radioactive decay
放射性损坏结构	metamictization
放射性碳计时	radiocarbon dating
放射性污染	radioactive pollution
放射性元素	radio-element
放射性元素矿床	ore deposit of radioactive elements
放射状	radiated
放射状集合体	radiating aggregate
飞来峰	klippe
非[寻]常光	extraordinary ray
非爆炸性喷发	non-explosive eruption
非地带性	azonality
非定向附生	nonepitaxial
非对称	asymmetry
非化学计量性	nonstoichiometry
非金属光泽	non-metallic luster
非金属矿产	nonmetallic ore
非金属矿床	nonmetallic deposit
非金属矿床学	geology of nonmetallic deposit
非金属矿物	non-metallic mineral
非晶化	non-crystallizing
非晶体	non-crystal
非晶质	amorphous substance
非晶质的	amorphous
非均键结构	anisodesmic structure
非均匀变形	inhomogeneous deformation

非均匀成核	heterogeneous nucleation
非均质的	inhomogeneous
非均质体	anisotropic body
非生物沉积构造	inorganic sedimentary structures
绯红	crimson, bright red
菲律宾	Philippines
霏细结构	felsitic texture
肥煤	fat coal
斐济	Fiji
翡翠	chalchuite
沸石化	zeolitization
沸石矿床	zeolite deposit
沸石水	zeolitic water
沸石相	zeolite facies
费尔斯曼	Fersman
镄	Fermium(Fm)
分层	lamination
分带序列	zoning sequence
分度圆筒与天平	graduated cylinder and balance
分解	disintegration, dissosiation
分解代谢作用	catabolism
分解者(体)	decomposer
分离结晶作用	fractional crystallization(FC)
分馏机理	fractionation mechanism
F-格子	F-lattice , face-centered lattice
分馏曲线	fractionation curve
分馏系数	fractionation factor
分馏效应	fractionation effect
分泌体	secretion
分凝作用	segregation
分配系数	partition coefficient, distribution coefficient
分散	dispersion

分散场	dispersion field
分散流	dispersion train
分散系数	dispersion coefficient
分散元素	dispersed element
分散元素矿床	ore deposit of dispersed elements
分散晕	dispersion halo
分选[作用]	sorting
分异指数	differentiation index(DI)
分域采样	stratigraphical sampling
分枝混合岩	ramification migmatite
分子轨道理论	molecular orbital theory
分子筛	molecular sieve
分子有机地球化学	molecular organic geochemistry
芬兰	Finland
玢岩	porphyrite
玢岩铁矿床	porphyrite iron deposit, porphyritic iron deposit
焚风	foehn wind
粉红色	pink, incarnadine, incarnate
粉末衍射标准卡片(JCPDS 卡)	JCPDS card
粉末状集合体	powdery aggregate
粉砂	silt
粉砂岩	siltstone
粪球粒	pellet
粪团粒	fecal pellet
丰度	abundance
风[力]选	winnowing
风暴	wind storm
风暴潮	surge
风暴流	storm current
风成的	aeolian
风成交错层理	wind-dune cross-bedding
风城	wind-eroded castle

风化层	regolith
风化带	weathering zone
风化壳	weathering crust
风化壳矿床	residual deposit, weathering crust mineral deposit
风化壳离子吸附型稀土矿床	weathering crust ion-adsorbed REE deposit
风化壳型硅酸镍矿床	nickeliferous silicate deposit in weathering crust
风化壳型铝土矿矿床	weathering-crust bauxite deposit
风化壳型锰矿床	manganese deposit in weathering crust
风化矿床	mineral deposit by weathering
风化作用	weathering
风积地貌	wind-accumulated landform
风积物	aeolian sediment
风棱石	ventifact
风沙流	wind-drift sand
风沙作用	wind-sand action
风蚀	wind erosion
风蚀残丘	deflation unaka
风蚀谷	aeolian valley
风蚀盆地	aeolian basin
风蚀洼地	blowout
风速	wind speed
风险评价	risk assessment
风向	wind direction
封闭	seal
封闭温度	closure temperature
封闭系统	closed system
锋面	front
蜂巢状构造	honeycomb structure
蜂窝状构造	honeycomb structure
蜂窝状网格	boxwork
缝合[线]	suture [line], stylolite

缝合带	suture zone
缝合面	stylolite
佛得角	Cape Verde
佛罗里达州(美)	Florida
佛蒙特州(美)	Vermont
弗吉尼亚州(美)	Virginia
弗氏旋转台	universal stage, Fedorov stage
芙蓉铀矿	furongite
氟	Fluorine(F)
氟磷灰石	fluorapatite
氟碳铈矿	bastnaesite
氟中毒	fluoride poisoning, fluorosis
浮生	overgrowth ,epitaxial growth
浮石	pumice
浮岩	pumice
浮游的	pelagic
符山石	vesuvianite, idocrase
幅合	convergence
幅散	divergence
辐射	radiation
辐射平衡	radiation balance
辐射热发光	radiothermoluminescence
斧石	axinite
斧石化	axinitization
俯冲	subduction
俯冲断层	underthrust fault
腐泥	sapropel
腐泥褐煤	saprodite
腐泥化作用	putrefaction, saprofication
腐泥煤	sapropelite
腐泥无烟煤	sapanthracite
腐泥烟煤	sapanthracon
腐植(殖)腐泥煤	humosapropelic coal

腐植(殖)煤	humulite
腐植组(褐煤显微组分组)	huminite
腐殖质	humus
负鲕	negative ooid
负光性	negative character
负荷压力	lithostatic pressure
负晶	negative crystal
负突起	negative relief
负形	negative form
负延性	length fast, negative elongation
负异常	negative anomaly
负载	load
附着能	attachment kinetics
复背斜	anticlinorium
复变质岩	polymetamorphic rock
复成分砾岩	polymict conglomerate
复成分粘土岩	polygenic claystone
复成矿床	polygenetic and compound deposit
复鲕	synthesize ooid, compound ooid
复合火山	compound volcano, composite volcano
复合双晶	compound twin(twinning)
复矿岩	polymineralic rock
复理石	flysch
复六方单锥	dihexagonal pyramid
复六方双锥	dihexagonal bipyramid
复六方柱	dihexagonal prism
复三方单锥	ditrigonal pyramid
复三方偏三角面体	ditrigonal scalenohedron
复三方双锥	ditrigonal bipyramid
复三方柱	ditrigonal prism
复式系统	multisystem
复四方单锥	ditetragonal bipyramid, ditetragonal pyramid

复四方偏三角面体	ditetragonal scalenohedron
复四方柱	ditetragonal prism
复稀金矿	polycrase
复向斜	synclinorium
复原重建	palinspastic reconstruction
复柱	biprism
副变基性岩	parametabasite
副变质岩	parametamorphic rock, parametamorphite
副角闪岩	paraamphibolite
副矿物	accessory mineral
副片麻岩	paragneiss
副片岩	paraschist
副热带高压	subtropical high
副像	paramorphism
傅立叶合成	Fourier synthesis
傅立叶转换红外光谱	Fourier Transform Infrared Spectroscopy (FTIS)
傅立叶转换拉曼光谱	Fourier Transform Raman Spectroscopy (FTRS)
富钙铝包体	calcium-aluminum-rich inclusion
富钙无球粒陨石	calcium-rich achondrite
富集地幔Ⅰ	enriched mantle Ⅰ(EM Ⅰ)
富集地幔Ⅱ	enriched mantle Ⅱ(EM Ⅱ)
富集系数	concentration coefficient
富集作用	concentration
富金线	pay streak
富矿体	ore shoot
富勒烯(C60 晶体)	fullerite, fullerence
富营养化	eutrophication
富铀岩体	uranium-rich massif
虎眼	chatoyancy

G

伽马射线光谱	Gamma-Ray Spectroscopy (GRS)

钆	Gadolinium(Gd)
噶尔本结构(以柱状矿物定向分布于片理面上为特点)	garben texture
改造	alteration
钙	Calcium(Ca)
钙长辉长无球粒陨石,钙长辉长岩	eucrite
钙长石	anorthite
钙钒石榴子石	goldmanite
钙沸石	scolecite
钙锆石榴子石	kimzeyite
钙铬石榴子石	uvarovite
钙硅酸盐角岩	calc-silicate hornfels
钙硅酸盐粒岩	calc-silicate fels
钙硅酸盐岩	calc-silicate rock
钙碱性系列	calc-alkaline series
钙碱指数	calc-alkali index
钙铝黄长石	gehlenite
钙铝石榴子石	grossularite,grossular
钙镁橄榄石	monticellite
钙十字沸石	phillipsite
钙钛矿	perovskite
钙钛矿-钛磁铁矿矿床	perovakite-titanomagnetite deposit
钙铁辉石	hedenbergite
钙铁石榴子石	andradite
钙霞石	cancrinite
钙质	calcareous
钙质板岩	calc-slate
钙质长石砂岩	calcareous feldspathic arenite
钙质片麻岩	calc-gneiss,calcareous gneiss
钙质片岩	calc-schist
钙质千枚岩	calc-phyllite
钙质岩	calc rock
钙柱石	meionite

盖玻片	cover slip
盖层	cover strata, overburden
盖层系	calymmian
盖雅假说	Gaia hypothesis
干谷	dry valley
干旱	drought
干酪根	kerogen
干气	dry gas, lean gas
干涉(扰)波痕	interference ripple
干涉色	interference colors
干涉色级序	order of interference color
干涉图	convergent light figure, interference figure
干盐湖	playa
干燥指数	aridity index
杆沸石	thomsonite
橄长岩	troctolite
橄榄[绿]色	olive green, olivaceous, olive
橄榄古铜球粒陨石	olivine-bronzite chondrite
橄榄石	olivine
橄榄石的蛇纹石假象	pseudomorphic of serpentine after olivine
橄榄铜矿	olivenite
橄榄岩	peridotite
橄榄陨铁	pallasite
橄榄紫苏球粒陨石	olivine-hypersthene chondrite
赣南矿	gananite
冈比亚	Gambia
冈瓦纳古大陆	Gondwanaland
刚果	Congo
刚玉	corundum
港湾结构	embayed texture
高 U/Pb 值的地幔(HIMU)	high m value(HIMU)
高场强元素	high field strength element (HFSE)

高点	culmination
高电压电子显微镜	High-Voltage Electron Microscopy (HVEM)
高度效应	altitude effect
高分辨场发射扫描电子显微镜	High-Resolution Field-Emission Scanning Electron Microscopy(HR-FE-SEM)
高分辨程序控制 X 射线层析	High-Resolution Computed X-ray Tomography (HRCXT)
高分辨电子显微镜	High-Resolution Electron Microscopy (HREM)
高分辨热重分析	High-Resolution Thermogravimetric Analysis (HRTGA)
高分辨同步辐射 X 射线粉末衍射	High-Resolution Synchrotron X-ray Powder Diffraction (HRSXPD)
高分辨透射电子显微镜	High-Resolution Transmission Electron Microscopy (HRTEM)
高分辨中子粉末衍射	High-Resolution Neutron Powder Diffraction (HRNPD)
高级白	high white
高级变质岩	high-grade metamorphic rock
高级片麻岩	high-grade gneiss
高岭石	kaolinite
高岭土化	kaolinization
高岭土矿床	kaolin deposit
高铝矿物	high aluminum mineral
高铝玄武岩	high-alumina basalt
高镁方解石	high-Mg calcite
高热变质[作用]	pyrometamorphism
高热变质岩	pyrometamorphic rock
高热交代矿床	pyrometasomatic deposit
高热交代岩	pyrometasomatic rock
高突起	high relief

高温钠长石	analbite
高温热液矿床	hypothermal deposit
高温水热矿床	hypothermal deposit
高压	high pressure
高压变质带	high pressure metamorphic belt
高压变质岩	high-pressure metamorphic rock
高压差热分析	High-Pressure Differential Thermal Analysis (HP-DAT)
高压釜	autoclave
高压相系	high-pressure facies series
高原	plateau
高原冰川	plateau glacier
锆	Zirconium(Zr)
锆[英]石	zircon
锆矿床	zirconium deposit
戈壁	gobi
戈尔德施密特	Goldschmidt V M
戈尔德施密特规则	Goldschmidt's rule
哥伦比亚	Colombia
哥伦比亚特区(美)	District of Columbia
哥斯达黎加	Costa Rica
格林纳达	Grenada
格林威治时间	Greenwich time
格陵兰(丹)	Greenland
格鲁吉亚	Georgia
格子[状]双晶	crosshatch twinning, grid twin, tartan twin
格子采样	cell sampling
格子结构	tartan structure
格子双晶	grid twin(tartan twin)
格子状结构	grating texture
格子状双晶	crosshatch twinning
隔热隔音材料	thermal and sound insulating materials

镉	Cadmium(Cd)
镉矿床	cadmium deposit
各向同性	isotropic
各向异性	anisotropy
铬	Chromium(Cr)
铬钾矿	lopezite
铬尖晶石	chrome-spinel
铬矿床	chromium deposit
铬铅矿	crocoite
铬酸盐	chromate
铬铁尖晶石	picotite
铬铁矿	chromite
铬铁矿浆	chromite ore magma
铬透辉石	chrome-diopside
根劈作用	splitting of root
根土岩	root clay
根源	provenance
更新世(统)	Pleistocene Epoch(Series)
更新统	Pleistocene
工具痕	tool marks
工业矿物	industrial mineral
工业矿物和岩石	industrial minerals and rocks
工艺美术原料矿产	mineral material for artware
工艺岩石学	technological petrology
公制的	metric
公转	revolution
汞	Mercury(Hg)
汞矿床	mercury deposit
汞铅矿	leadamalgam
共轭节理	conjugate joint
共格面网	common net
共价	covalency

共焦扫描光学显微镜	Confocal Scanning Optical Microscopy (CSOM)
共结晶	cotectic crystallization
共生	paragenesis
共生分析	paragenesis analysis
共生矿	paragenic ore
共生矿物	paragenetic mineral
共生组合	paragenetic assemblage
共型	cotype
共振离子化质谱仪	Resonant Ionization Mass-Spectrometry (RIMS)
构造	structure
构造背景	tectonic setting
构造超压	tectonic overpressure
构造窗	window, fenster
构造等值(高)线	structure contour
构造地貌	tectonic landform, structural landform
构造地球化学	tectono-geochemistry
构造地震	tectonic earthquake
构造地质学	structural geology
构造分析	structural analysis
构造高点	culmination
构造后变晶	post-tectonic blast
构造湖	tectonic lake
构造角砾岩	tectonic breccia
构造节理	tectonic joint
构造砾岩	tectonic conglomerate
构造盆地	tectonic basin
构造前变晶	pre-tectonic blast
构造圈闭	structural trap
构造山系	structural mountain
构造体制	tectonic regime
构造图	structural map

构造形迹	structural feature
构造旋回	tectonic cycle
构造学	tectonics
构造岩	tectonite
构造岩石学	structural petrology
构造运动	tectonization
构造资料	structural data
孤立组分	indifferent components
古巴	Cuba
古北矿	gupaiite
古成体	paleosome
古登堡不连续面	Gutenberg discontinuity
古地磁[学]	paleomagnetism
古地磁场	paleomagnetic field
古地磁极	paleomagnetic pole
古地磁条带	paleomagnetic belt
古地理	paleogeography
古地温	paleogeotemperature
古风化壳	paleocrust of weathering
古构造运动	paleotectonic movement
古近系	Palaeogene
古流分析	palaeocurrent analysis
古生代(界)	Palaeozoic Era(Erathem)
古生界	Palaeozoic(Pz)
古太古界	Paleoarchean
古铜钙长无球粒陨石	howardite
古铜橄榄石铁陨石	bronzite-olivine stony-iron
古铜辉石	bronzite
古铜-鳞英石铁陨石	siderophyre
古铜色	bronze
古土壤	palaeosol
古新世(统)	Paleocene Epoch(Series)
古新统	Paleocene

古元古代(界)	Palaeoproterozoic Era(Erathem)
古元古界	Palaeoproterozoic
古陨击坑	astrobleme
古紫色	antique violet
骨骼的	skeletal
骨架的	skeletal
骨痛病	itai-itai disease
钴	Cobalt(Co)
钴蓝色	cobalt blue,Hungary blue
鼓丘	drumlin
固定论	fixism
固定组分	fixed components
固-固反应	solid-solid reaction
固结系	Statherian
固溶体	solid solution
固溶体分解曲线	solvus exsolution curve
固溶体离溶(固溶体分解,出溶)	exsolution
固态化学	solid state chemistry
固体潮	earth tide,solid tide
固相线	solidus
顾家石	gugiaite
瓜德鲁普统	Guadalupian
瓜德罗普岛(法)	Guadeloupe
关岛(美)	Guam
冠状体	corona
贯穿双晶	penetrate twin
贯入矿体	injected ore body
贯入混合岩	injection migmatite
贯入片麻岩	injection gneiss
光电子光谱	Photo Electron Spectroscopy (PES)
光电子光谱仪	Photoelectron Spectromicroscopy (PESM)
光电子能损光谱	Photo Electron Energy-Loss Spectroscopy (PEELS)

光化学烟雾	photochemical smog
光亮煤	bright coal
光卤石	carnallite
光卤石矿床	carnallite deposit
光率体	indicatrix
光片	polished section
光性方位	optic orientation
光性符号	optical sign
光性异常	optical anormaly
光学发射光谱	Optical Emission Spectroscopy (OES)
光学扫描显微镜	Optical Scanning Microscopy (OSM)
光学石英矿床	optical quartz deposit
光学显微镜	optical microscopy (OM)
光学性质	optical property
光学原料矿产	raw material for optical use
光学主轴	principal optic axis
光泽	luster
光致发光	photoluminescence
光轴	optic axis,optic-axis(OA)
光轴角	optic-axial angle
光轴角色散	dispersion of optic angle
光轴面	optic-axial plane
光子发射光谱	Photoemission Spectroscopy (XPES)
圭亚那	Guyana
硅	Silicon(Si)
硅[质碎]屑的	siliciclastic
硅钙石	rankinite
硅华	geyserite
硅化	silicification
硅化带型铀矿	silicified zone type U-ore
硅灰石	wollastonite
硅灰石矿床	wollastonite deposit
硅碱指数	silic-alkali index

硅孔雀石	chrysocolla
硅铝层	sial
硅铝矿物	salic mineral
硅镁层	sima
硅镁石	humite
硅硼钙石	datolite
硅铍石化	bertranditization
硅酸的	silicic
硅酸盐	silicate
硅锌矿	willemite
硅藻类	diatom
硅藻土矿床	diatomite deposit
硅藻岩	diatomite
硅质	silicic
硅质的	silicic
硅质碎屑的	siliciclastic
贵橄榄石	chrysolite
贵金属矿床	precious metal deposit
滚动背斜构造	rollover anticline structure
国际符号(也称 HM 符号)	international symbol
过饱和度	supersaturation
过饱和岩	oversaturated rock
过渡带	transitional zone
过渡型键	transition bond
过渡型元素	transitional element，siderophile element
过渡性支撑	transitional support
过去全球变化	past global change
过剩氩	excess argon
过剩组分	excess components

H

铪	Hafnium(Hf)
铪矿床	hafnium deposit

哈勃定律	Hubble's law
哈默斯利型铁矿床	Hamersley-type iron deposit
哈萨克斯坦	Kazakhstan
骸晶	skeleton crystal
骸晶的	skeletal
骸晶结构	skeleton texture
骸晶形貌	skeletal morphology
海流	sea current,ocean current
海岸	coast
海百合[灰岩]	encrinite
海滨	seashore, seaside,beach
海滨砂矿	beach placer,sea-beach placer
海昌蓝	hydron blue
海底成矿作用	submarine metallization
海底黑烟柱	black smoker
海底扩张	sea floor spreading
海底锰结核	manganese nodules on the ocean floor, pelagite
海底喷发	submarine eruption
海底喷流作用	submarine exhalative process
海底热流	submarine heat flow,heat flow through the ocean floor
海底热泉	submarine hot spring
海底生物	benthonic
海地	Haiti
海沟	trench
海积地貌	sea deposit landform
海积阶地	marrine-built terrace, marrine-built platform
海积作用	sea deposit
海岬	cape
海进	transgression
海军蓝	navy blue

海蓝色	pale bluish green
海浪	sea wave
海浪冲刷	alluvion
海陆交替相沉积矿床	paralic sedimentary deposit
海绿色	sea green
海绿石	glauconite
海绿石石英砂岩	glauconitic quartzarenite
海绵[体]	sponge
海绵岩	spongolite
海绵陨铁结构	sideronitic texture
海绵状物	sponge
海泡石	sepiolite
海泡石矿床	sepiolite deposit
海泡石粘土矿床	sepiolite clay deposit
海平面变化	sea-level change
海平面变化的	eustatic
海平面气压	sea-level pressure
海穹	sea arch
海上油气田	offshore oil gas field
海蚀[作用]	wave erosion
海蚀凹槽	sea cave
海蚀地貌	sea erosion landform
海蚀洞(穴)	marine cave
海蚀阶地	marine-cut bench, rock bench
海蚀崖	sea cliff
海蚀柱	stack
海滩	beach
海滩岩	beachrock
海退	regression
海西期	Hercynian
海湾	bay, embayment
海湾状熔蚀	embayment
海相沉积磷块岩矿床	marine sedimentary phosphorite deposit

海相沉积铝土矿矿床	marine sedimentary bauxite deposit
海相沉积锰矿床	marine sedimentary manganese deposit
海啸	tsunami,seismic sea wave
海洋	ocean
海洋赤潮	ocean red tide
海洋地球化学	marine geochemistry
海洋化探	marine geochemical exploration
海洋环境	marine environment
海因里希事件	Heinlich event
氦	Helium(He)
含金石英脉	gold-quartz vein
含矿建造	ore formation
含煤岩系	coal-bearing series, coal measures
含煤岩系古地理	paleogeography of coal-bearing series
含煤岩系旋回结构	sedimentary cycle in coal-bearing series
含气饱和度	gas saturation
含水饱和度	water saturation
含水层	aquifer
含水熔融曲线	hydrous melting curve
含铜砂岩	cupreous sandstone
含铜页岩型矿床	kupferschiefer type deposit
含氧系数	oxygen coefficient
含油饱和度	oil saturation
含油地层	oil-bearing formation
含油面积	oil-bearing area
含油气构造	oil gas structure
含油气盆地	petroliferous basin
含油气区	petroleum region
含油气圈闭	oil gas trap
含油气省	petroleum province
含铀硅岩	uranium-bearing silicalite
含铀胶磷矿	uraniferous collophane
含铀页岩	uranium-bearing shale

含有化石的	fossiliferous
焓(热函)	Enthalpy (H)
寒武纪(系)	Cambrian Period(System), Cambrian
寒武纪生物爆发	Cambrian explosion
韩国	Republic of Korea
航空化探	airborne geochemical exploration
行列	row
合成矿物	synthetic mineral (artificial mineral)
合成作用	anabolism
合纹石	plessite
河床	(river) bed
河床砂矿	riverbed placer
河谷	valley
河谷阶地	valley terrace
河谷砂矿	vally placer
河控三角洲	river-dominated delta
河口沙坝	river mouth bar
河口湾	firth, river branching bay
河流	river
河流的	fluvial
河流地貌	river landform
河流回春	rejuvenation of stream
河流阶地	river terrace
河流学	rheology
河漫滩	washland
河滩	beach
核部	core
核磁共振光谱	Nuclear Magnetic Resonance Spectroscopy (NMRS)
核地球化学	nuclear geochemistry
核冬天	nuclear winter
核反应	nuclear reaction
核年代学	nucleochronology

核燃料废物	nuclear fuel waste
核酸-蛋白质	nucleic-acid-protein
核形石	oncolite
荷兰	Netherlands
荷属安的列斯	The Netherlands Antilles
褐黑色	brownish black
褐红色	maroon
褐-黄色多色性	pleochroism brown-yellow
褐帘石	allanite,orthite
褐绿色	brownish green
褐煤	lignite, brown coal
褐色	brown,filemot
褐铁矿	limonite
赫茨普龙-罗素图	H-R diagram
赫-罗图	H-R diagram
黑潮[日]	kuroshio
黑电气石	schorl
黑洞	black hole
黑金刚石	carbonado,carbon diamond
黑矿型矿床	kuroko deposit
黑硼锡镁矿	magnesiohulsite
黑色	black
黑色金属矿床	ferrous metal deposit
黑色冶金辅助原料矿床	raw material for ferrous metallurgy
黑钨矿	wolframite
黑钨矿-石英脉	wolframite-quartz vein
黑曜岩	obsidian
黑硬绿泥石	stilpnomelane
黑云母	biotite
黑云母化	biotitization
黑子周期	sunspot cycle
痕迹化石	ichnofossils
痕量元素	trace element

很低级变质岩	very low-grade metamorphic rock
很高突起	very high relief
恒定性资源	perpetual resource
恒星	star
恒星原子核合成	nucleosynthesis in star
横波(S 波)	transverse wave,S wave
横交结构	diablastic texture
横剖面	cross section
横切面	basal section
横弯褶皱	bending fold
横向冲击波	lateral blast
横向分带	transversal zoning
横向沙丘	transversal dune
横褶皱	cross fold
烘烤带	baked zone
红宝石	ruby
红海海底成矿作用	submarine metallization in the Red Sea
红褐色	henna,mahogany,rufous,russet,russety, bronze,reddish orange
红花黄	safflower yellow
红椒色	paprika
红巨星	red giant
红蓝宝石矿床	ruby and sapphire deposit
红蓝色	reddish blue
红帘石	piedmontite
红色	red
红色蚀变	red-color alteration
红闪石	katoforite,cataphorite
红砷镍矿	niccolite
红石矿	hongshiite
红土	red soil
红土化	lateritization
红土化型铁矿床	laterized iron deposit

红外光谱	Infrared Spectroscopy (IRS)
红移	red shift
红藻	red algae
红柱石	andalusite
红柱石-夕线石型(系列)	andalusite-sillimanite type(or series)
红棕色	reddish brown
洪都拉斯	Honduras
洪积砂矿	proluvial placer
洪积扇	proluvial fan
洪水	flood
后成合晶	symplectite
后成河	subsequent river
后继形貌	epimorphism
后生矿床	epigenetic deposit
后生异常	epigenetic anomaly
厚板状	tabular
厚层白云岩化	massive dolomitization
弧-沟间隙	arc-trench gap
弧后盆地	back-arc basin
弧前盆地	fore-arc basin
胡桃色	nut-brown
湖滨	beach
湖泊	lake
湖泊的	lacustrine
琥珀[黄]色	amber
互层	alteration
互代性	diadochy
互生	intergrowth
花斑岩	granophyre
花岗变晶多边形结构	granoblastic-polygonal texture
花岗变晶多缝合结构	granoblastic-polysutured texture
花岗变晶结构	granoblastic texture
花岗结构	granitic texture

花岗闪长岩	granodiorite
花岗岩	granite
花岗岩化[作用]	granitization
花岗岩化成矿说	granitization theory of metallization
花岗岩类	granitoid
花岗岩型铀矿	granite type U-ore
花岗质混合片麻岩	granitic migmatitic gneiss
花岗质片麻岩	granitic gneiss
花呢结构	tweed texture
花青	flower blue
华盛顿州(美)	Washington
华夏古大陆	Cathaysia
滑动反应	sliding reaction
滑距	slip
滑坡	landslide,earth slide
滑石	talc
滑石矿床	talc deposit
滑石菱镁[片]岩	listvenite
滑石片岩	talc schist
滑塌	slump
滑移[作用]	glide
滑移面	glide reflection plane
滑移双晶	glide twin,mechanical twin
化肥原料矿产	fertilizer minerals
化工和化肥原料	chemical and fertilizer raw materials
化工原料矿产	mineral material for chemical industry
化石燃料	fossil fuels
化石相	biofacies
化学沉积构造	chemical sedimentary structures
化学沉积矿床	mineral deposit by chemical sedimentation
化学沉积岩	chemical sedimentary rock
化学地球动力学	chemical geodynamics
化学动力学	chemical kinetics

化学反应制动原理	chemical reaction brake principle
化学风化	chemical weathering
化学计量学	stoichiometry
化学位	chemical potential
化学物种	chemical species
化学压实	chemical compaction
怀俄明州(美)	Wyoming
还原	reduce
还原环境	reducing environment
环斑结构	rapakivi texture
环礁湖	lagoon
环境保护	environment protection
环境背景[值]	environmental background
环境变迁	environmental transition
环境储库	environmental reservoir
环境地球化学	environmental geochemistry
环境地球化学参数	environmental geochemistry parameter
环境规划	environmental planning
环境化学演化	chemical evolution of environment
环境介质	environmental medium
环境界面	environmental interface
环境界面地球化学	environmental interface geochemistry
环境模拟	environmental simulation
环境年代学	environmental chronology
环境区划	environmental regionalization
环境退化	environmental degradation
环境退化造成的资源短缺	resource scarcity due to environmental degradation
环境污染	environmental pollution
环境物质迁移	transport of environmental substance
环境物质释放	release of environmental substance
环境物质形态	form of environmental substance
环境物质转化	transformation of environmental substance

环境效应	environmental effect
环境异常	environmental anomaly
环境因子	environmental factor
环境影响	environmental influence
环境影响评价	environmental impact assessment
环境有机地球化学	environmental organic geochemistry
环境质量	environmental quality
环境质量变异	environmental quality variation
环境质量标准	environmental quality standard
环境质量参数	environmental quality parameter
环境质量评价	environmental quality assessment
环境质量图	environmental quality atlas
环境质量指数	environmental quality index
环境自净	environmental self-purification
环太平洋成矿带	circum-Pasific metallogenic belt
环形山	crater
环状硅酸盐	cyclosilicate(ring silicate)
环状双晶	ring-shaped twin,cyclic twin
缓冲	buffering
缓冲反应	buffer reaction
缓冲剂	buffering agent
缓冲作用	buffer action
缓流	tranquil flow
幻数规则	magic number rule
荒漠化	desertification
荒漠区	desert area
黄白色	yellowish white
黄长石	melilite
黄河石	huanghoite
黄褐色	khaki,snuff color
黄钾铁矾	jarosite
黄色	yellow,yellowness
黄砷榴石	berzeliite

黄水晶	citrine
黄锑矿	cervantite
黄铁绢英岩	beresite
黄铁绢英岩化	beresitization
黄铁矿	pyrite
黄铁矿化	pyritization
黄铁矿矿床	pyrite deposit
黄铜矿	chalcopyrite
黄土	loess
黄土地貌	loess landform
黄土高原	loess plateau
黄土梁	loess liang
黄土峁	loess mao
黄土塬	loess yuan
黄玉	topaz
煌斑结构	lamprophyric texture
煌斑岩	lamprophyre
灰白色	off-white,offwhite,hoar
灰硅钙石	spurrite
灰褐色	taupe,grayish brown
灰化	ashing
灰黄色	grayish yellow
灰蓝色	stone blue
灰色	gray,grey
灰质簇	sclerodermite
灰棕色	grayish brown
恢复	recovery
挥发性元素	volatile elements:
辉铋矿	bismuthinite
辉长结构	gabbro texture
辉长岩	gabbro
辉沸石	stilbite
辉绿结构	diabasic texture,ophiolite texture

辉绿岩	diabase,dolerite
辉钼矿	molybdenite
辉钼矿-石英脉	molybdenite-quartz vein
辉砷钴矿	cobaltite
辉石	pyroxene
辉石角岩	pyroxene hornfels
辉石角岩相	pyroxene-hornfels facies
辉石片麻岩	pyroxene gneiss
辉石岩	pyroxenite
辉锑矿	antimonite,stibnite
辉铜矿	chalcocite
辉岩	pyroxenolite
辉银矿	argentite
汇聚型板块边界	convergent plate boundary
汇聚型板块边缘	convergent plate margin
会聚光	convergent
会聚束电子衍射	Convergent-Beam Electron Diffraction (CBED)
彗星	comet
混合白云石化[作用]	dorag dolostone
混合花岗岩	migmatitic granite
混合晶体	mixed crystal
混合模型	mixing model
混合片麻岩	migmatitic gneiss
混合驱动储油层	combination drive reservoir
混合水	mixed water
混合水白云岩	dorag dolostone
混合岩	migmatite
混合岩化[作用]	migmatism,migmatization
混合岩化变质岩	migmatization metamorphic rock
混合岩化作用	migmatization
混合岩浆	ichor
混合杂岩	migmatic complex

混积岩	diamictite
混染岩	hybridrock
混杂堆积	melange
混杂陆源沉积岩	diamictite
混杂岩	mélange
混浊性	turbidity
活动带	mobile belt
活动构造	active tectonics
活动构造带	active tectonic belt
活动论	mobilism
活动型造山作用	activation-type orogeny
活动元素吸收系数	mobile element absorption coefficient (MAC)
活动组分	mobile components
活化转移	remobilization
活化转移成矿作用	ore-forming process by remobilization
活火山	active volcano
活塞-缸筒设备	piston-cylinder apparatus
活性酸度	active acidity
活性碳法	absorbent charcoal method
活性铀	mobile uranium
火成岩	igneous rock
火山	volcano
火山玻璃	[volcanic] glass
火山玻屑	glass shard,vitric pyroclast
火山沉积型矿床	volcano-sedimentary deposit
火山沉积型硼矿床	volcano-sedimentary boron deposit
火山成矿作用	volcanogenic ore-forming process
火山成因块状硫化物矿床	volcanogenic massive sulfide deposit
火山成因矿床	volcanogenic mineral deposit
火山带	volcanic belt
火山弹	[volcanic] bomb
火山地貌	volcanic landform

火山地震	volcanic earthquake
火山构造	volcanic structure
火山-构造拗陷	volcano-tectonic depression
火山管	diatreme
火山弧	volcanic arc
火山弧岩石	volcanic arc rock
火山灰	[volcanic] ash
火山灰凝灰岩	lapilli tuff
火山活动	volcanic activity
火山机构(体)	volcanic edifice, volcanic apparatus
火山晶屑	crystal pyroclast
火山颈	volcanic neck
火山口	crater
火山口湖	caldera lake, crater lake
火山砾	lapilli
火山砾岩	lapillistone
火山泥[石]流	lahar, volcanic mudflow
火山喷[火]口	volcanic vent
火山穹隆	volcanic dome
火山砂	volcanic sand
火山碎屑	pyroblast, tephra
火山碎屑的	pyroclastic
火山碎屑降落	tephra fall
火山碎屑结构	pyroclastic texture
火山碎屑流	pyroclastic flow, ash flow
火山碎屑岩	pyroclastic rock
火山碎屑锥	pyroclastic cone, tephra cone
火山通道	volcanic feeder, [volcanic] conduit, volcanic vent
火山旋回	volcanic cycle
火山学	volcanology
火山岩	volcanic rock

火山岩黄铁矿型多金属矿床	pyritiferous polymetallic deposit in volcanic rocks
火山岩黄铁矿型铜矿床	pyritiferous copper deposit in volcanic rocks
火山岩相	volcanic facies
火山岩屑	lithic pyroclast
火山岩型铀矿	volcanic type U-ore
火山韵律	volcanic rhythm
火山渣	cinder
火山渣锥	cinder cone, scoria cone
火山锥	volcanic cone
火山作用	volcanism
火焰熔化测试	Flame and Fusion Testing (FFT)
火焰状构造	flame structure
火灾	conflagration
钬	Holmium(Ho)

I

I-格子	I-lattice, body-centered lattice

J

击变玻璃	diaplectic glass
击象	impact marks, percussion marks
饥荒	famine
机械沉积矿床	mineral deposit by mechanical sedimentation
机械风化	mechanical weathering
机械双晶	mechanical twin
鸡窝状矿体	nest-like ore body
基本粒子	fundamental particle
基本烈度	fundamental intensity
基底	basement
基底成熟度	basement maturity
基底式胶结	basal cementation
基里巴斯	Kiribati
基鲁纳铁矿床	Kiruna iron deposit

基体	substrate, substratum
基性	basic
基性角岩	basic hornfels
基性麻粒岩	basic granulite
基性岩	basic rock, basite
基岩	bedrock
基岩海岸	bedrock coast
基质	groundmass, matrix
基质支撑	matrix-support
基准井	key well, stratigraphic well
畸变	distortion
激光氟化同位素分析	Laser-Fluorination Isotope Analysis (LFIA)
激光干涉显微镜	Laser-Interference Microscopy (LIM)
激光-拉曼探针光谱	Laser-Raman Microprobe Spectroscopy (LRMS)
激光离子化表面分析	Surface Analysis By Laser Ionization (SALI)
激光磨蚀硫同位素探针光谱	Laser-Ablation Sulfur Isotope Microprobe Spectroscopy(LASIM)
激光磨蚀硫同位素显微镜	Laser-Ablation Sulfur Isotope Microscopy (LASIM)
激光磨蚀探针分析	Laser-Ablation Microprobe Analysis (LAM)
激光散射分析	Laser Scattering (LS)
激光扫描光学显微镜	Laser Scanning Optical Microscopy (LSOM)
激光探针分析	Laser-Microprobe Analysis (LMA)
激光同位素分析	Laser Isotope Analysis (LIA)
激光诱发荧光光谱	Laser-Induced Fluorescence Spectroscopy (LIFS)
激浪带	surf zone
吉布斯自由能	Gibbs free energy

吉布提	Djibouti
吉尔吉斯斯坦	Kirgizstan
级粒沉积	graded sediment
极不完全解理	cleavage in trace
极粗砂岩	very coarse sandstone
极化	polarization
极距角	pole distance angle
极射赤平投影	stereographic projection
极完全解理	eminent cleavage, perfect cleavage
极细砂岩	very fine sandstone
极性反转	polarity-reversal
极移	polar wandering
极游移轨迹	polar wandering path
极圆的	well rounded
极震区	meizoseismal area
极轴	polar axis
急流	torrent flow
集合体	aggregate
集块岩	agglomerate
集料	aggregate
集群绝灭	mass extinction
几何单形	geometric form
几何结晶学	geometrical crystallography
几内亚	Guinea
几内亚比绍	Guinea-Bissau
脊线	crest line
计时	age dating
计算机模拟	computer model
纪	Period
季风	monsoon
季节性吸收系数	temporal absorption coefficient(TAC)
继承氩	inherited argon
寄生火山锥	parasitic cone

寄生褶皱	parasitic fold
蓟县矿	jixianite
加成组分	additive components
加大边结构	enlargement texture
加积[作用]	aggradation
加里东期	Caledonian
加利福尼亚(美)	California
加拿大	Canada
加拿大树胶	Canada balsam
加纳	Ghana
加篷	Gabon
加速质谱仪	Accelerated Mass-Spectroscopy (AMS)
夹层	intercalation
夹矸	parting, dirt band
夹石	horse-stone, dunn bass
镓	Gallium(Ga)
镓矿床	gallium deposit
岬角	cape
钾	Potassium(K)
钾长片麻岩	kali-feldspar gneiss
钾长石化	potash feldspathization
钾化带	potassic zone
钾交代[作用]	potassic metasomatism
钾交代型铀矿	potassic-metasomatism type uranium deposit
钾芒硝矿床	glaserite deposit
钾霞石	kaliophilite
钾硝石	niter
钾-氩计时	K-Ar dating
钾盐	sylvite, sylvine
钾盐矿床	sylvite deposit, potash salt deposit
价带	valence band
架状硅酸盐	framework silicate(tectosilicates)

假白榴石	pseudoleucite
假次生包体	pseudo-secondary inclusion
假等变线	false isograd
假等时线	pseudoisochron
假晶	crystal cast, pseudomorphic crystal
假蓝宝石	sapphirine
假熔岩	buchite
假色	pseudochromatic color
假团粒	pseudopeloid
假象	pseudomorph, pseudomorphism
假象的	pseudomorphic, pseudomorphous
假玄武玻璃	pseudotachylite
假异常	false anomaly
假整合	para-unconformity, disconformity
尖晶石	spinel
尖晶石矿床	spinel deposit
尖棱褶皱	chevron fold
尖灭的	tapered
尖-圆形褶皱	cuspate-lobate folds
间冰期	interglacial stage
间层混合岩	lit-par-lit migmatite
间层片麻岩	lit-par-lit gneiss
间断	hiatus
间隔劈理	spaced cleavage
间粒结构	intergranular texture
间歇分带	intermittent zoning, pulsative zoning
间歇熔融	batch melting
间隐结构	tholeiitic texture
柬埔寨	Kampuchea
减少	reduce
减灾	natural hazard reduction
剪节理	shear joint
剪切波(S 波)	shear wave (S wave)

剪切褶皱	shear fold
剪应力	shear stress
简并系统	degenerated system
简单的脱水反应	simple dehydration reaction
简单的脱碳酸反应	simple decarbonation reaction
简化	reduce
碱	alkalis
碱长片麻岩	alkali-feldspar gneiss
碱湖	natron lake
碱交代[作用]	alkaline metasomatism
碱交代型铀矿	alkalic-metasomatism type uranium deposit
碱土	alkaline earths
碱性长石	alkali feldspar
碱性环境	alkaline environment
碱性辉长岩	essexite
碱性辉石化	alkaline pyroxenization
碱性角闪石化	alkaline amphibolization
碱性系列	alkaline series
碱性玄武岩	alkali basalt
碱性岩	alkali rock
碱性岩型铀矿	alkaline rock type uranium deposit
碱玄岩	tephrite
建筑材料	constructional materials, building materials
建筑材料矿产	raw material for building and construction
健康效应	health effect
渐(递)进变形	progressive deformation
渐新世(统)	Oligocene Epoch(Series)
溅落堤	spatter rampart
溅落锥	spatter cone
鉴定矿物学	determinative mineralogy
键长	bond length
键能	bond energy
键强	bond strength

姜黄色	ginger
降尘	dustfall
降解	degradation
降解作用	degradation
降水量	amount of precipitation
绛色	deep red, crimson
交叉结构	decussate texture
交错层理	cross-bedding
交错纹层	cross-lamination
交代	replacement
交代[作用]	metasomatism, replacement, metasomatosis
交代变质[作用]	metasomatic metamorphism
交代残余结构	metasomatic relict texture
交代晶体结构	metacrystal texture
交代矿床	metasomatic mineral deposit
交代岩	metasomatic rock
交沸石	harmotome
交换反应	exchange reaction
交换平衡	exchange equilibrium
交换组分	exchange components
交面线理	intersecton lineation
交生	intergrowth
交替	alteration
交织河流	braided stream
交织结构	pilotaxitic texture
胶结[作用]	cementation
胶结类型	cementation type
胶结物	cement
胶结物的世代	cement generation
胶粒	colloid
胶磷矿	cellophane
胶泥煤	saprocollite
胶体	colloid

胶体沉积	colloidal deposition
胶体化学沉积矿床	mineral deposit by colloidal agglutination
胶体结晶作用	collocrystallization
胶体矿物	colloidal mineral
胶状构造	colloform structure
焦力天平	Jolly Balance (JB)
焦煤	coking coal
礁圈闭[构造]	reef trap[structure]
角斑岩	keratophyre
角度不整合	angular unconformity
角度分辨 X 射线光电子谱仪	Angle-Resolved X-ray Photoelectron Spectroscopy(ARXPS)
角峰	horn
角砾岩	breccia
角砾状构造	breccia structure
角砾状混合岩	agmatite
角闪片岩	hornblende schist
角闪石	amphibole
角闪石片麻岩	hornblende gneiss
角闪石岩	hornblendite, hornblende rock
角闪岩	amphibolite
角闪岩相	amphibolite facies
角岩	hornfels
角银矿	chlorargyrite
角质体	cutinite
搅动	agitate
搅动浸出	agitated leaching
阶	stage, module
阶地	terrace
阶地砂矿	bench placer
阶段加温	stepwise heating
阶梯状断层	step fault
接触变质岩	contact metamorphic rock

接触-交代变质[作用]	contact-metasomatic metamorphism
接触交代矿床	contact-metasomatic deposit
接触交代岩	contact metasomatic rock
接触模式原子力显微镜	Contacting Mode Atomic Force Microscopy (CMAFM)
接触-热变质[作用]	contact-thermal metamorphism
接触热变质岩	contact thermometamorphic rock
接触-热变质作用	contact-thermal metamorphism
接触式胶结	contact cementation
接触双晶	contact twin
接触晕	contact aureole
节理	joint
节理面	joint surface
节理系	joint system
节理组	joint set
结点	node
结构	texture
结构成熟度	textural maturity
结构倒转	texture reverse
结构腐植体	humotelinite
结构基元	motif
结构镜质体	telinite
结构水	constitution water
结构退变	texture reverse
结构因子	structure factor
结合水	bound water
结核	nodule
结核体	concretion
结晶白云岩	crystalline dolomite
结晶单形	crystallographic form
结晶度	crystallinity
结晶分异矿床	crystallization-differentiation deposit
结晶符号	crystal indices

结晶灰岩	crystalline limestone
结晶结构	crystalline texture
结晶片岩	crystalline schist
结晶水	crystallization water
结晶体	crystal
结晶习性	crystal habit
结晶学	crystallography
结晶岩	crystalline rock
结晶指数	crystallization index
结晶质	crystalline substance
结晶轴	crystallographic axis
结晶作用	crystallization
捷克	Czech Rep.
解理	cleavage
解理面	cleavage plane
介电常数	dielectric constant
介电性	dielectricity
介形虫	ostracode
介质	agent
界	Erathem
界限粘土	boundary clay
金	Gold(Au)
金[黄]色	golden [yellow]
金伯利岩	kimberlite
金刚光泽	adamantine luster
金刚石	diamond
金刚石矿床	diamond deposit
金刚石压腔	diamond anvil cell
金红石	rutile
金矿床	gold deposit
金沙江石	jinshajiangite
金-铀砾岩	auri-uraniferous conglomerate
金云母	phlogopite

金属光泽	metallic luster
金属互化物	intermetallic compound ,alloy
金属矿产	metallic ore
金属矿床	ore deposit,metallic mineral deposit
金属矿物	metallic mineral
金属镍铁	metal nickel-iron
金属有机矿化作用	metallo-organic matter mineralization
金属有机络合物	metallo-organic complex
金字塔形沙丘	pyramidal dune
津巴布韦	Zimbabwe
紧密堆积	closer packed
堇青石	cordierite
近变质岩	anchi-metamorphic rock
近滨[的]	nearshore
近海的	hemipelagic
近海型含煤岩系	paralic coal-bearing series
近海岩	hemipelagite
近火山活动金属矿床	proximal ore deposit
进变质	prograde metamorphism
进变质岩	prograde metamorphic rock
进化论	evolution theory
进浪流	beach drift
浸出	leaching
浸染状构造	disseminated structure
禁带	forbidden band
经度	longitude
经济地质学	economic geology
经济矿床	economic mineral deposit
晶胞	[unit] cell
晶胞参数	cell parameter
晶畴	domain
晶簇	druse
晶簇构造	drusy structure

晶带	[crystal] zone
晶带定律	zone law
晶带符号	zone symbol
晶带指数	zone index
晶带轴	zone axis
晶洞的	vuggy
晶洞构造	miarolitic structure
晶格	lattice
晶格点	lattice point
晶格能	lattice energy
晶格同位	diadochy
晶格位能	lattice site energy
晶格优选方位	lattice preferred orientation
晶核	crystal nucleus
晶化	crystallizing,devitrification
晶界塑性变形	intercrystalline plastic deformation
晶类	crystal class
晶棱	crystal edge
晶面	crystal face
晶面条纹	striation,striae
晶面指数	index of crystal face
晶内塑性变形	intracrystalline plastic deformation
晶体	crystal
晶体测量	goniometry
晶体常数	crystal constants
晶体场理论	crystal field theory(CFT)
晶体定向	crystal orientation
晶体对称定律	law of crystal symmetry
晶体符号	crystal indices
晶体光学	crystal optics
晶体化学	crystallochemistry
晶体化学式	crystallochemical formula
晶体结构学	crystallology

晶体缺陷	crystal defect
晶体生长	crystal growth
晶体物理	crystal physics
晶系	crystal system
晶腺	geode
晶屑	crystal pyroclast
晶形	crystal form
晶芽	crystal nucleus
晶质	crystalline substance
晶质铀矿	uraninite
晶族	crystal category
景观地球化学	landcape geochemistry
净初级生产量	net primary production
净砂岩	arenite
净转移反应	net-transfer reaction
径流	runoff
静海石	tranquillityite
静海相的	euxinic
静砂岩	arenite
静水压力	hydrostatic pressure
静态重结晶	static recrystallization
镜煤	vitrain
镜质组	vitrinite
韭闪石	pargasite
臼齿构造	molar tooth (MT) structures
就地	situ
居里温度	Curie temperature
居里-吴里弗原理	Curie-Wulff theory
居里原理	Curie theory
居留时间	residence time
锔	Curium(Cm)
局部背景	local background
局部变质[作用]	local metamorphism

局部化学的	topochemical
局部平衡	local equilibrium
桔红色	nacarat
桔黄色	saffron [yellow], bisque
橘红色	jacinth
橘黄色	croci, saffron yellow
橘色	orange
矩阵	matrix
巨晶	megacryst
巨砾	boulder
巨行星	giant planet
飓风	hurricane
锯齿状断口	hackly fracture
聚沉[作用]	flocculation
聚合模型	polymerization model
聚合物	polymer
聚合作用	polymerization
聚集元素	aggregated element
聚晶的	polycrystalline
聚敛板块边界	convergent plate boundary
聚煤盆地	coal basin
聚煤期	coal-forming period
聚煤区	coal accumulating region
聚煤作用	coal accumulation
聚片双晶	multiple twinning, polysynthetic twin
聚形	combination form
聚形纹	combination striations
卷曲结构	cylindrical and helicitical
卷状铀矿体	uranium roll
绢英带	phyllic zone
绢云母	sericite
绢云母化	sericitization
绝对丰度	absolute abundance

绝灭	extinction
绝灭背景质	background extinction
绝热压缩	adiabatic compression
绝缘材料矿产	mineral material for insulation
绝缘体	insulator
均变论	uniformitarianism
均衡	isostasy
均衡说	isostasy theory
均衡异常	isostatic anomaly
均键结构	isodesmic structure
均量(衡)补偿	isostatic compensation
均斜	homocline
均一的	homogeneous
均一法	homogenization method, homogeneity
均夷[作用]	gradation
均匀变形	homogeneous deformation
均匀层理	homogeneous bedding
均匀成核	homogeneous nucleation
均匀的	homogeneous, homogenetic, homogenic
均质的	homogeneous, homogenetic, homogenic
均质混合岩	homogenic migmatite
均质体	isotropic body
菌类体	sclerotinite
菌体	thallus

K

咖啡色	tawny, dark brown, tan, coffee
喀麦隆	Cameroon
喀斯特	karst
喀斯特地貌	karst landform, karst topography
喀斯特作用	karstification, karst process
卡尔斯巴双晶	Carlsbad twin
卡拉哈里锰矿床	Karahari manganese deposit

卡腊贾斯型铁矿床	Calacas-type iron deposit
卡林型金矿床	Carlin-type gold deposit
卡姆巴尔达镍矿床	Kambalda nickel deposit
卡-钠复合双晶	Carlsbad-albite compound twin
卡其色	khaki
卡式双晶	Carlsbad twin
卡塔尔	Qatar
开发井	development well
开放系统	open system
开曼群岛(英)	Cayman Islands
开形	open form
锎	Californium(Cf)
勘查地球化学	exploration geochemistry
勘查地质学	exploration and prospecting geology
堪萨斯州(美)	Kansas
康拉德不连续面	Conrad discontinuity
康涅狄格州(美)	Connecticut
抗磁性	diamagnetism
钪	Scandium(Sc)
柯石英	coesite
科里奥利力	Coriolis force
科罗拉多州(美)	Colorado
科马提岩	komatiite
科摩罗	Comoros
科特迪瓦	Cote d'Ivoire
科威特	Kuwait
颗[形]石	coccolith
颗粒	grain
颗粒灰岩	grainstone
颗粒结构	grained texture
颗粒流	grain flows
颗粒泥晶灰岩	wackestone
颗粒支撑	grain-support

颗石鞭毛藻	coccolithophora
颗石藻	coccolith
壳层规则	shell rule
壳源[的]	crust-derived
壳质组	exinite
可采资源	recoverable resource
可持续发展	sustainable development
可持续利用	sustainable use
可分裂的	fissile
可见与近红外反射光谱	Visible and Near-Infrared Reflectance Spectroscopy (VNIRS)
可见宇宙	the visible universe
可逆的(双向的)转变	enantiotropic transformation
可燃性	combustibility
可容纳空间	accommodation
可压缩性	compressibility
克拉克	Clarke F W
克拉克值	Clarke,clark value
克拉通	craton
克拉通内的	intracratonic
克里沃罗格式铁矿床	Krivoyrog-type iron deposit
克列里奇液	Clerici's solution
克罗地亚	Croatia
克山病	Keshan disease
刻赤式铁矿床	Kerch iron deposit
氪	Krypton(Kr)
肯尼亚	Kenya
肯皮尔赛铬矿床	Kempirsay chromium deposit
肯塔基州(美)	Kentucky
空间化学	space chemistry
空间群	space group
空晶	negacrysts
空晶石	chiastolite

空气对比比重仪(比重瓶)	Air Comparison Pycnometer
空缺	vacancy
空隙	interstice, void
空隙的	interstitial
孔雀蓝	peacock blue
孔雀石	malachite
孔隙	pore
孔隙度	porosity
孔隙式胶结	porous cementation
孔隙水	porosity water
孔兹岩	khondalite
枯叶色	filemot
苦橄岩	picrite
苦橄质	picritic
库尔斯克铁矿床	Kursk-type iron deposit
库克群岛(新)	Cook Islands
块滑石	soapstone
块体运动	mass movement
块状白云岩化	massive dolomitization
块状层理	massive bedding
块状构造	massive structure
块状集合体	massive aggregate
矿层	ore bed, reef, seam,
矿产	ore, useful mineral, mineral resource
矿产经济学	mineral economics
矿产开发	mineral exploitation
矿产勘查	mineral exploration
矿产普查	mineral prospecting
矿巢	ore nest
矿床	mineral deposit, ore deposit
矿床[成因]学	metallogeny
矿床[地质]学	geology of ore deposits
矿床成因类型	genetic type of mineral deposits

矿床成因说	metallogeny
矿床地球化学	mineral deposit geochemistry
矿床地质学	mineral deposit geology
矿床分带	zonal distribution of mineral deposits
矿床工业类型	commercial type of mineral deposits
矿床模式	mineral deposit model
矿床系列	mineragenetic series of ore deposits
矿床学	study of ore deposit, mineral deposits
矿床组合	association of ore deposits
矿带	ore zone
矿点	ore occurrence
矿顶相	facies of ore body top
矿根相	facies of ore body root
矿化集中(金属)区学说	theory of concentration of metallogenesis
矿化剂	mineralizer
矿化流体	mineralizing fluid
矿化作用	mineralization
矿浆	ore magma
矿浆型铁矿床	ore magma iron deposit
矿脉	vein, lode
矿帽	capping
矿棉	mineral wool
矿囊	ore pocket
矿区	ore district
矿山	mine
矿山地质学	mining geology
矿石	ore
矿石构造	structure of ores
矿石建造	ore formation
矿石结构	texture of ores
矿石矿物	ore mineral
矿石矿物学	ore mineralogy
矿石品级	grade of ore

矿石品位	tenor of ore
矿石学	ore petrology
矿石自然类型	natural type of ore
矿胎	protore
矿体	ore body
矿体产状	mode of occurrence of ore body, attitude of ore body
矿体形态	morphology of ore body
矿体形状	form of ore body
矿田	ore field
矿筒	ore pipe, ore chimney
矿物	mineral
矿物成熟度	mineralogical maturity
矿物等时线	mineral isochron
矿物共生[组合]	mineral paragenesis
矿物合成	mineral synthesis
矿物化学	mineral chemistry
矿物化学式	mineral formula
矿物模式分析	Mineralogical Model Analysis (MMA)
矿物物理	mineral physics
矿物线理	mineral lineation
矿物相变	phase transformation of mineral
矿物相律	mineralogical phase rule
矿物形态	morphology of minerals
矿物学	mineralogy
矿物颜色	color of mineral
矿物组合	mineral association, mineral assemblage
矿相显微镜	ore microscope
矿-岩时差	rock-ore formation time interval
矿源层	source bed
矿源岩	source rock
矿柱	ore column, ore pillar
亏损的 MORB 地幔	depleted MORB mantle(DMM)

亏损地幔	depleted mantle(DM)
魁北克省(加)	Quebec
扩散	diffusion
扩散分馏	diffusional fractionation
扩散交代[作用]	diffusive metasomatism
扩散流动	diffusive flow
扩散晕	diffusion halo
扩散作用	diffusion
扩展X射线能失精细结构光谱	Extended X-ray Energy-Loss Fine-Structure Spectroscopy (EXELFS)
扩展X射线吸收精细结构光谱	Extended X-ray Absorption Fine-Structure Spectroscopy (EXAFS)

L

拉斑[玄武]结构	intergranular texture
拉斑系列	tholeiitic series
拉斑玄武岩	tholeiite
拉长石	labradorite
拉德洛统	Ludlow
拉曼光谱	Raman Spectroscopy (RS)
拉尼娜现象	La Nina
拉伸系	Tonian
拉伸线理	stretching lineation
拉脱维亚	Latvia
蜡状光泽	waxy luster
莱茨-杰利折射计	Leitz-Jelley refractor
莱河矿	Laihunite
莱索托	Lesotho
铼	Rhenium(Re)
铼矿床	rhenium deposit
兰多弗里统	Llandovery
兰多理论	Landau
蓝透闪石	Winchite

蓝宝石	sapphire,sappire
蓝方石	haüyne
蓝黑色	blue black
蓝晶石	kyanite,disthene
蓝晶石片岩	kyanite schist
蓝晶石-夕线石型(系列)	kyanite-sillimanite type(or series)
蓝绿色	aquamarine blue,cyan,bluish green
蓝片岩	blue schist,blueschist
蓝片岩相	blueschist facies
蓝色	blue
蓝闪石	glaucophane
蓝闪石片岩	glaucophane schist
蓝石棉矿床	crocidolite deposit,crocidolite-asbestos deposit
蓝铁矿	vivianite
蓝铜矿	azurite
蓝线石	dumortierite
蓝藻细菌	cyanobacteria
镧	Lanthanum(La)
郎道理论	Landau
浪控三角洲	wave-dominated delta
劳厄群	Laue group
劳伦古大陆	Laurentia
劳亚古大陆	Laurasia
铹	Lawrencium(Lr)
老第三纪(系)	Paleogene Period(System)
老化	ageing
老挝	Laos
铑	Rhodium(Rh)
乐平煤	Loping coal, lopite
乐平统	Lopingian
勒辛型铀矿床	Rössing-type uranium deposit
雷公墨	tektite

镭	Radium(Ra)
类(矿物分类单位)	type
类地行星	terrestrial planet,earth-like planet
类结晶的(结晶程度不好的)	paracrystalline(poorly crystalline)
类木行星	Jovian planet
类质同象混合物	isomorphic mixture
类质同象组分	isomorphic components
类质同象	isomorphism,allomerism
累乘晕	multiplicative halo
累积系数	accumulative coefficient
累积因子	accumulative factor
累积作用	accumulative action
累加晕	additive halo
棱角状的	angular
冷封口高压容器	cold seal pressure vessel
冷凝带	chill zone
冷提取	cold extraction
离散[型]板块边界	divergent plate boundary
离散型板块边缘	divergent plate boundary
离子交换	ion exchange
离子性	ionicity
黎巴嫩	Lebanon
理论海平面	ecstatic sea level
理论资源	speculative resource
锂	Lithium(Li)
锂电气石	elbaite
锂辉石	spodumene
锂矿床	lithium deposit
锂铍石	liberite
锂云母	lepidolite
力	force
力学性质	mechanical property
立方晶系	cubic system,isometric system

立方体	cube
立方最紧密堆积	cubic closest packing
立陶宛	Lithuania
立体图	three dimentional diagram,block diagram
立体显微镜	Stereoscopic Microscopy (SM)
利比里亚	Liberia
利比亚	Libya
利蛇纹石	lizardite
沥青	bitumen, asphalt
沥青光泽	pitchy luster
沥青砂	tar sand
沥青铀矿	pitchblende
沥青质体	bituminite
沥青质页岩	bituminous shale
栗色	nut-brown,maroon
砾[石]	gravel,pebble
砾岩	conglomerate
砾岩型铀矿	conglomerate type U-ore
粒度递变沉积	graded sediment
粒硅钙石	tilleyite
粒硅镁石	chondrodite
粒级	grain size
粒级层理	graded bedding
粒间的	interparticle
粒屑灰岩	grainstone
粒序层理	graded bedding
粗玄岩	dolerite
粒雪	firn snow
粒岩	fels
粒岩或××岩	fels or rock
粒状	granular
粒状灰岩	grainstone

粒子诱导 X 射线发射光谱	Particle-Induced X-ray Emission Spectroscopy (PIXE)
连岛沙坝	tie bar
连线转换反应	tieline-switching reaction
连续	continuity
连续不等粒斑状结构	seriate porphyritic texture
连续反应	continuous reaction
连续劈理	continuous cleavage
连续相变	continuous transformation
链状硅酸盐	inosilicate(chain silicate)
两栖类	amphibia
亮红色	vivid red
亮晶	sparry
亮晶胶结物	sparry cement
亮煤	clarain
钌	Ruthenium(Ru)
列支敦士登	Liechtenstein
烈度	intensity
裂变径迹保留年龄	fission-track retention age
裂变径迹计时	fission-track dating
裂缝	fracture
裂缝矿脉	gash ore body
裂谷	rift
裂开	parting
裂开面	parting plane
裂理	parting
裂隙	fissure
裂隙脉	fissure vein
裂隙喷发	fissure eruption
裂隙式喷发	fissure eruption
裂隙水	crevice water, fracture water
裂陷	rift
鬣刺结构	spinifex texture

林伍德	Ringwood A E
林伍德规则	Ringwood's rule
临界半径	critical radius
临界矿物组合	critical mineral assemblage
临界浓度	critical concentration
临界资源	critical resource
淋积矿床	leaching deposit
淋积型铀矿	leaching type uranium deposit
淋滤	leaching
淋滤矿床	leaching deposit
磷	phosphorus(P)
磷光	phosphorescence
磷灰石	apatite
磷灰石矿床	apatite deposit
磷灰石岩	apatitolite
磷块岩矿床	phosphorite deposit
磷矿床	phosphorous deposit(phosphate deposit)
磷酸盐	phosphate
磷酸盐矿床	phosphate deposit
磷霞岩	urtite
磷钇矿	xenotime
鳞绿泥石	thuringite
鳞片变晶结构	lepidoblastic texture
鳞片状的	scaly
鳞石英	tridymite
灵敏高质量分辨离子探针	Sensitive High-Mass Resolution Ion Microprobe (SHRIMP)
陵夷[作用]	degradation
菱沸石	chabazite
菱镁矿	magnesite
菱镁矿矿床	magnesite deposit
菱锰矿	rhodochrosite
菱面体	rhombohedron

菱面体晶系	rhombohedral system, trigonal crystal system
菱铁矿	siderite
菱锌矿	smithsonite
菱形十二面体	dodecahedron, rhombic dodecahedron, rhombododecahedron
流变学	rheology
流动构造	flow structure, fluidal structure
流痕	current ripple, tool marks
流面构造	planar flow structure
流水地貌	fluvial landform
流水作用	fluvial action
流体包裹体	fluid inclusion
流体包裹体地质温度显微镜	Fluid Inclusion Geothermometric Microscopy(FIGM)
流体包裹体显微温度分析	Fluid Inclusion Microthermometric Analysis(FIMA)
流体超压	fluid overpressure
流体动力学	fluid dynamics
流体对流	fluid convection
流体压力	fluid pressure
流体-岩石交互作用	fluid-rock interaction
流体粘度	fluid viscosity
流纹构造	fluidal structure
流纹结构	rhyotaxitic texture
流纹岩	rhyolite
流纹质	rhyolitic
流线构造	linear flow structure
流星	meteor
流行的	prevalent
流行地幔	prevalent mantle(PREMA)
流域	river basin
流状构造	fluxion structure
留尼汪岛(法)	Reunion

硫化物	sulphide
硫化物矿床次生富集带	secondary enriched zone of sulfide deposits
硫化物矿床次生富集作用	secondary enrichment of sulfide deposits
硫化物矿床氧化带	oxidized zone of sulfide deposit
硫矿床	sulphur deposit
硫砷钌矿	ruarsite
硫酸盐	sulphate
硫循环	sulfur cycle
榴辉岩	eclogite
榴辉岩相	eclogite facies
瘤状构造	nodular structure
锍	Sulphur(S)
六八面体	hexakisoctahedron, hexoctahedron
六方单锥	hexagonal pyramid
六方钾霞石	kalsilite
六方晶系	hexagonal system
六方偏方面体	hexagonal trapezohedron
六方双锥	hexagonal bipyramid
六方异极晶体	hexagonal hemimorphic class
六方柱	hexagonal prism
六方最紧密堆积	hexagonal closest packing, hexagonal close-packed
六面体	hexahedron
龙卷风	tornado
漏斗晶	hopper crystal
露	dew
露头	outcrop
卢森堡	Luxembourg
卢旺达	Rwanda
卤水	brine
卤水矿床	brine deposit
卤族元素	halogen family element
镥	Lutetium(Lu)

陆表海	epeiric sea
陆成(生)的	terrigenous
陆间裂谷	intercontinental rift
陆-陆碰撞	continent-continent collision
陆上油气田	onshore oil gas field
陆下岩石圈	subcontinental lithosphere
陆源的	terrigenous
陆源区	provenance
陆源碎屑	terrigenous detrital
陆源碎屑沉积物	terrigenous clastic sediment
陆源碎屑物	terrigenous detrital sediment
陆源碎屑岩	terrigenous clastic rock
鹿毛色	fawn
路易斯安那州(美)	Louisiana
铝	Aluminium(Al)
铝硅酸盐	aluminosilicate
铝矿床	aluminium deposit
铝石榴子石	sudoite
铝土矿	bauxite
铝土矿矿床	bauxite deposit
铝直闪石	jedrite
铝质岩	aluminous rocks
绿钙闪石	hastingsite
绿辉石	omphacite
绿蓝色	greenish blue
绿帘角闪岩	epidote amphibolite
绿帘-角闪岩相	epidote-amphibolite facies
绿帘石	epidote
绿帘石化	epidotization
绿帘石岩	epidotite
绿硫钒脉状矿床	patronite lode deposit
绿泥石	chlorite
绿泥石化	chloritization

绿泥石型铀矿	U-ore of chlorite type
绿片岩	green schist, greenschist
绿片岩相	greenschist facies
绿色	green
绿松石	turquoise
绿松石矿床	turquoise deposit
绿松石色	turquoise blue
绿脱石	nontronite
绿纤石	pumpellyite
绿岩	greenstone
绿岩带	greenstone belt
绿藻	green algae
绿柱石	beryl
绿柱石化	berylitization
绿柱石石英脉	beryl-quartz vein
绿柱石伟晶岩矿床	beryl pegmatite deposit
氯	Chlorine(Cl)
氯磷灰石	chlorapatite
滦河矿	luanheite
略带紫色的	purplish, purply
轮式双晶	cyclic twin
罗得岛州(美)	Rhode Island
罗马尼亚	Romania
罗盘	compass
螺距	screw distance
螺旋生长台阶	spiral steps
螺旋轴	screw rotation axis
裸子植物	gymnosperm
裸子植物时代	age of gymnosperm
洛林式铁矿床	Lorraine-type iron deposit

M

麻粒岩	granulite

麻粒岩相	granulite facies
马鞍状	saddle-shaped
马达加斯加	Madagascar
马尔代夫	Maldives
马耳他	Malta
马拉维	Malawi
马来西亚	Malaysia
马里	Mali
马里兰州(美)	Maryland
马尼托巴省(加)	Manitoba
马其顿	Macedonia
马萨诸塞州(美)	Massachusetts
马绍尔群岛共和国	Republic of the Marshall Islands
马提尼克岛(法)	Martinique
马尾丝状构造	horsetail structure
埋藏变质[作用]	burial metamorphism
埋藏变质岩	burial metamorphic rock
埋藏阶地	buried terrace
埋藏矿体	buried ore body
脉	vein
脉动分带	pulsative zoning, intermittent zoning
脉动说	pulsation theory
脉混合岩	arterite
脉融合岩	venite
脉石	gangue
脉石矿物	gangue mineral
脉体	vein material
脉状层理	flaser bedding
脉状构造	vein structure
脉状铜矿床	vein-type copper deposit
曼尼巴双晶	Manebach twin
幔源[的]	mantle-derived

漫反射红外傅立叶转换光谱	Diffuse Reflectance Infrared Fourier Transfer Spectroscopy (DRIFT)
芒特艾萨多金属矿床	Mount Isa polymetallic deposit
芒硝	mirabilite
芒硝矿床	mirabilite deposit
盲谷	blind valley
盲矿	blind ore
盲矿体	blind ore body
猫眼	chatoyancy
毛发状集合体	capillary aggregate
毛里求斯	Mauritius
毛里塔尼亚	Mauritania
毛细管浓缩	capillary concentration
毛细水	capillary water
玫瑰[红]色	rose color, rosy, rosiness
媒介物	agent
煤[炭]	coal
煤变质作用	coal metamorphism
煤层	coal seam
煤层对比	correlation of coal seam
煤层分叉	bifurcation of coal seam
煤沉积模式	sedimantary model of coal
煤成气	coal gas
煤成岩作用	coal diagenesis
煤地球化学	coal geochemistry
煤垩	smut
煤核	coal ball
煤华	blossom
煤化作用	coalification
煤级	coal rank
煤精	jet
煤矿地质	coal mining geology
煤盆地	coal basin

煤田	coalfield
煤[田]地质学	coal geology
煤系	coal-bearing series, coal measures
煤相	coal facies
煤岩学	coal petrology, coal petrography
镅	Americium(Am)
美利坚合众国	United States of America
美属维尔京群岛	Virgin Islands of the United States
镁	Magnesium(Mg)
镁电气石	dravite
镁橄榄石	forsterite
镁矿床	magnesium deposit
镁铝石榴子石	pyrope
镁钠闪石	magnesioriebeckite
镁闪石	magnesiocummingtonite
镁铁[质]岩	mafic rock
镁铁闪石	cummingtonite
镁铁质	mafic
镁铁质岩	mafic rock
镁星叶石	magnesioastrophyllite
镁叶云母	eastonite
镁质	magnesian
镁质角岩	magnesian hornfels
镁质岩[石]	magnesian rock
钔	Mendelevium(Md)
蒙大拿州(美)	Montana
蒙古	Mongolia
蒙钠长石	monalbite
蒙特塞拉特岛(英)	Montserrat
蒙脱石	montmorillonite, smectite
蒙脱石化	montmorillonitization
锰	Manganese(Mn)
锰矿床	manganese deposit

锰铝石榴子石	spessartite, spessartine
锰帽	manganese hat
锰帽型锰矿床	manganese-hat-type manganese deposit
孟加拉国	Bangladesh
糜棱结构	mylonitic texture
糜棱岩	mylonite
米兰科维奇学说	Milankovich theory
米勒-尤里反应	Miller-Urey reaction
米勒指数	Miller's indices
米纳斯吉拉斯型铁矿床	Minas Gerais-type iron deposit
米色	beige
米氏符号	Miller's indices
米制的	metric
秘鲁	Peru
密度	density
密度补偿	density compensation
密度测试	Density Determination (DD)
密度流	density current
密封[墙]	seal
密克罗尼西亚联邦	The Federated States of Micronesia
密苏里州(美)	Missouri
密西西比河谷式铅锌矿床	Mississippi-valley[-type] lead-zinc deposit
密西西比亚系	Mississippian
密西西比州(美)	Mississippi
密歇根州(美)	Michigan
蜜黄长石	meliphanite
缅甸	Burma
缅因州(美)	Maine
面角	interfacial angle
面角守恒定律	law of constancy angle
面金属量	areal productivity
面理	foliation
面网	net

面网间距	interplanar spacing
面网密度	reticular density
面向	younging [direction]
面心格子	face-centered lattice, F-lattice
明矾石	alunite
明矾石化	alunitization
明矾石矿床	alunite deposit
明亮的	bright, lightful
明尼苏达州(美)	Minnesota
冥古宙	Hadean
模	Module
模拟	simulation
模拟实验	simulation experiment
模式年龄	model age
摩檫发光	triboluminescence
摩尔多瓦	Moldova
摩尔体积计算	Molar Volume Calculation (MV)
摩洛哥	Morocco
摩纳哥	Monaco
摩氏硬度	Mohs' hardness
摩氏硬度计	Mohs' hardness scale (MHS)
磨[片]	grind
磨拉石	molasses
磨蚀[作用]	abrasion
磨圆的	rounded
磨圆度	roundness
蘑菇石	mushroom
末元古系	Neoproterozoic Ⅲ
莫霍洛维奇不连续面	Mohorovicic discontinuity (Moho)
莫霍面	Mohorovicic discontinuity (Moho)
莫来石	Mullite
莫桑比克	Mozambique
莫托沙拉锰矿床	Mutusala manganese deposit

墨黑的	jet-black, jetty, pitch-black, coal-black, jet
墨晶	black quartz
墨西哥	Mexico
默硅钙镁石	merwinite
母群	supergroup
母岩	mother rock
木栓质体	suberinite
钼	Molybdenum(Mo)
钼矿床	molybdenum deposit
钼铅矿	wulfenite
钼酸盐	molybdenate
幕	episode
穆斯保尔光谱	Mossbauer Spectroscopy (MBS)

N

镎	Neptunium(Np)
内[碎]屑	intraclast
内布拉斯加州(美)	Nebraska
内部等时线	internal isochron
内部自动平衡	homeostasis
内动力	endogenetic force
内反射	internal reflection
内核	inner core
内华达州(美)	Nevada
内缓冲	internal buffering
内缓冲组分	internally buffered components
内加热高压容器	internally heated pressure vessel
内陆盆地	inland basin, interior basin
内陆型含煤岩系	limnic coal-bearing series
内生成矿作用	endongenic metallization
内生矿床	endogenic mineral deposit
内生异常	endogenetic anomaly
内生作用	endogenic process, endogenesis

内碎屑亮晶灰岩	intrasparite
内碎屑泥晶灰岩	intramicrite
内乌斯塔德双晶	Neustad twin
内营力	endogenetic force
纳米比亚	Namibia
钠	Sodium(Na)
钠钡长石	banalsite
钠长-绿帘角岩	albite-epidote hornfels
钠长-绿帘角岩相	albite-epidote hornfels facies
钠长片麻岩	albite gneiss
钠长石	albite
钠长石化	albitization
钠长石律双晶	albite law
钠沸石	natrolite
钠交代[作用]	sodic metasomatism
钠交代型铀矿	sodic-metasomatism type uranium deposit
钠锰电气石	tsilaisite
钠闪石	riebeckite
钠铁闪石(又称:亚铁钠闪石)	arfvedsonite
钠透闪石	richterite
钠硝石	soda niter,nitronatrite,nitratine
钠硝石矿床	soda niter deposit, nitratine deposit
钠黝帘石化	saussuritization
钠云母	paragonite
钠柱石	marialite
氖	Neon(Ne)
耐火材料	refractory materials,refractory
耐火粘土矿床	fire clay deposit
南达科他州(美)	South Dakota
南非	South Africa
南卡罗来纳州(美)	South Carolina
南平石	nanpingite
南斯拉夫	Yugoslavia

难熔的	refractory
挠曲	flexure
挠性	flexibility
瑙鲁	Nauru
嫩绿	viridescence
能带理论	band theory
能散 X 射线谱仪	Energy-Dispersive X-ray Spectroscopy (EDX, EDXS)
能[量色]散谱仪	Energy-Dispersive Spectroscopy (EDS)
尼泊尔	Nepal
尼加拉瓜	Nicaragua
尼科波尔矿床	Nikopol manganese deposit
尼科尔棱镜	polarizing prism
尼日尔	Niger
尼日利亚	Nigeria
泥[石]流	mudflow
泥[状]岩	mudrock,argillite
泥板岩	argillite,shale
泥化	argillization
泥灰岩	marl
泥火山	mud volcano,mudlump
泥晶灰岩(曾用名:微晶灰岩)	micrite
泥晶颗粒灰岩	packstone
泥砾	boulder-clay
泥粒[状]灰岩	packstone
泥裂	mud crack
泥盆纪(系)	Devonian Period(System),Devonian
泥丘	mudlump
泥石流	debris flow
泥滩	mud bank
泥炭	peat
泥炭化作用	peat formation
泥铁矿	ironstone

泥质	pelitic
泥质角岩	pelitic hornfels
泥质岩	argillite, pelite
泥质岩石	pelitic rocks
泥状结构	muddy texture
铌	Niobium(Nb)
铌钽矿床	niobium and tantalum deposit
铌钽伟晶岩矿床	niobium-tantalum pegmatite deposit
霓辉石	aegirine-augite
霓石	aegirine
霓石化	aegrinization
霓霞岩	ijolite
拟晶	crystalline mimetic
逆冲断层系	thrust system
逆磁性	diamagnetism
逆断层	reverse fault
逆温	temperature inversion
逆向分带	reverse zoning
逆掩断层	overthrust fault
年代	Age
年代地层单位	chronostratigraphic unit
粘度	viscosity
年龄测定	age determination
年龄谱	age spectrum
粘土化	clayization
粘土化蚀变型铀矿	argillified type U-ore
粘[滞]性	viscosity
粘土矿床	clay deposit
粘土矿物	clay mineral
粘土岩	argillite
粘土质	argillaceous
粘滞性	tenacity
鸟粪石磷矿床	guano-type phosphate deposit

鸟眼构造	birdseye structure
镍	Nickel(Ni)
镍黄铁矿	pentlandite
镍矿床	nickel deposit
镍纹石	taenite
柠檬[黄]色	lemon, lemon-yellow, citrine, citreous
凝灰结构	tuff texture, ash texture
凝灰岩	(lapilli) tuff
凝胶化作用	gelification
凝结	condensation
凝聚	flocculation, condensation
凝块[石]	clot
凝析气田	condensate field
凝析油	condensate oil
牛轭湖	oxbow lake
扭力天平	torsion balance
扭折	kinking, kink
扭折带	kink band
纽埃(新)	Niue
纽伯格蓝	Neuberg blue
纽芬兰省(加)	Newfoundland
纽约州(美)	New York
浓度克拉克值	Clarke of concentration
浓度梯度	concentration gradient
浓集系数	concentration coefficient
浓集中心	concentration center
钕	Neodymium(Nd)
暖池	warm pool
挪威	Norway
诺马尔斯基干涉相衬图	Nomarski Interference Contrast Imaging (NICI)
诺依曼线	Neumann line
锘	Nobelium(No)

O

欧拉极	Evler pole
偶奇效应	even-odd mass effect
偶数规则	even rule
耦合反应	coupled reaction

P

爬迹	trails
爬行动物时代	reptilian age
爬行类	reptilia
帕特森图	Patterson diagram
拍岸浪	breaker
抛光	polish
刨蚀[作用]	glacial ploughing
PDB 标准(碳同位素标准,美国南卡罗纳白垩系皮迪组的美洲似箭石)	PDB standard(Pee Dee belemnite standard)
培长石	bytownite
锫	Berkelium(Bk)
配分粒序	distribution grading
配位数	coordination number (CN)
喷出口	vent
喷出相	extrusive facies
喷出岩	extrusive rock
喷发柱	eruption column
喷流矿床	exhalation deposit
喷气矿床	exhalation deposit
盆地	basin
盆-岭构造	basin-range structure
彭志忠石	Pengzhizhongite
硼	Boron(B)
硼矿床	boron deposit
硼镁石	szaibelyite

硼镁石矿床	ascharite deposit，szaibelyite deposit
硼镁铁矿	ludwigite
硼砂	borax
膨润土	bentonite
膨润土矿床	bentonite deposit
膨胀说	expansion theory
膨胀珍珠岩原料矿床	expanded perlite material deposit
碰撞型造山[作用]	collision-type orogeny
pH－Eh 图解（酸碱度与氧化还原电位图解）	pH-Eh diagram
批式熔融	batch melting
披覆构造	draping structure
铍	Beryllium(Be)
铍矿床	beryllium deposit
劈理	cleavage
劈理扇	cleavage fan
劈理域	cleavage domain
皮壳状构造	crusty structure
皮壳状脉	crustified vein
皮特凯恩群岛（英）	Pitcairn Islands
片	sheet
片沸石	beulandite
片理	schistosity
片麻理	gneissosity
片麻岩	gneiss
片麻状构造	gneissose structure
片岩	schist
片状的	schistic
片状构造	schistose structure
偏方复十二面体	diakisdodecahedron
偏方面体	trapezohedron
偏方形	deltoid
偏光	polarized light

偏光棱镜	polarizing prism
偏光显微镜	polarizing microscope, Polariscopy (PS)
偏三角面体	scalenohedron
偏十二面体	deltohedron
偏提取	partial extraction
偏心鲕	eccentric ooid
偏倚	bias
偏应力	deviatoric stress
偏振	polarization
偏振角	angle of polarization
漂白土	fuller's earth
漂白土矿床	fuller's earth deposit
漂尘	dust
漂砾	boulder
贫钙无球粒陨石	calcium-poor achondrite
贫煤	meager coal
品蓝	reddish blue
平底晶洞构造	stromatactis
平衡常数	equilibrium constant
平衡畴	equilibrium domain
平衡剖面	balanced section
平衡相图	equilibrium phase diagram
平均应力	mean stress
平流层	stratosphere
平面偏振	plane-polarization
平坦状断口	even fracture
平卧矿体	manto
平卧褶皱	recumbent fold
平行不整合	para-unconformity, disconformity
平行层理	parallel bedding
平行电子能失光谱	Parallel Electron Energy-Loss Spectroscopy (PEELS)
平行连晶	parallel grouping, parallel intergrowth

平行六面体	parallel hexahedron
平行双晶	parallel twin
平行双面	pinacoid
平行消光	parallel extinction, straight extinction
平行褶皱	parallel fold
平移断层	strike-slip fault
平移轴	translation axis
平原	plain
评价井	evaluation well, appraisal well
坪年龄	plateau age
苹果绿	apple green
钋	Polonium(Po)
坡积砂矿	deluvial placer
坡积铁矿床	slope-wash iron deposit
坡缕石	palygorskite
坡面	dome
钷	Promethium(Pm)
破火山口	caldera
破浪	breaker
破裂	fracture
破劈理	fracture cleavage
剖面	profile
$P-T-t$ 轨迹	P-T-t path
葡萄酒红	wine red
葡萄石	prehnite
葡萄石-绿纤石相	prehnite-pumpellyite facies
葡萄牙	Portugal
葡萄状构造	botryoidal structure
葡萄状集合体	botryoidal aggregate
镤	Protactinium(Pa)
普遍的	prevalent
普蓝	Berlin blue
普里多利统	Pridoli

普鲁士蓝	prussian blue, bronze blue
普通地质学	physical geology
普通辉石	augite
普通角闪石	hornblende
普通角闪石角岩	hornblende hornfels
普通角闪石角岩相	hornblende hornfels facies
普通铅	common lead
普通球粒陨石	ordinary chondrite
谱系采样	hierarchical sampling
镨	Praseodymium(Pr)

Q

期	stage, episode
奇点	singularity
骑田岭矿	Qitianlingite
起源	provenance
气藏	gas pool
气成-热液矿床	pneumato-hydrothermal deposit
气顶气驱储油层	gas cap drive reservoir
气候	climate
气候带	climatic belt, climatic zone
气候系统	climate system
气候型	climatic type
气孔构造	vesicular structure
气煤	gas coal
气苗	gas seepage
气泡	bubble
气体	gas
气体保留年龄	gas retention age
气体地球化学测量	geochemical gas survey
气田	gas field
气团	air mass
气温	air temperature

气温递减率	air temperature lapse rate
气旋	cyclone
气压	atmospheric pressure
气液变质岩	pneumato-hydrothermal metamorphic rock
气油比	gas-oil radio
契尔马克分子	tschermakite
器官系数	acropetal coefficient(AC)
恰图拉矿床	Chiatura manganese deposit
千枚岩	phyllite
千枚状构造	phyllitic structure
千糜岩	phyllonite
牵引负载	traction [load]
牵引流	tractive current
牵引总体	traction [load]
铅	Lead(Pb)
铅[灰]色	leaden
铅 207 富集指数	delta 7/4 =(^{207}Pb/^{204}Pb—^{207}Pb/^{204}PbNHRL) * 100, ^{207}Pb/^{204}PbNHRL = 15. 627 + 1. 209 (^{207}Pb/^{204}Pb))
铅白色	lead white
铅帽	leadhat
铅锌矿床	lead-zinc deposit
前寒武纪	Precambrian
前生期化学演化	chemical evolution in prebiological period
前震	fore-shock,foreshock
潜火山相	subvolcanic facies
潜火山岩	subvolcanic rock
潜山圈闭[构造]	buried-hill trap[structure]
潜蚀[作用]	suffusion
潜水	phreatic water,underground water, groundwater,ground water
潜穴	burrows

潜在酸度	potential acidity
浅[橄榄]绿色的	laurel-green
浅成高温热液矿床	xenothermal deposit
浅成高温水热矿床	xenothermal deposit
浅成侵入体	shallow intrusion
浅成侵入相	hypabyssal intrusive facies
浅成热液矿床	epithermal deposit
浅成岩	hypabyssal rock
浅带	epizone
浅粉红色	baby pink
浅海	shallow sea，epeiric sea
浅海相沉积矿床	neritic sedimentary deposit
浅褐色	sandy beige, beige
浅黑的	darkish, swarthy
浅红	pale red
浅黄	pale yellow
浅黄褐色	fawn
浅灰褐色	ficelle
浅灰蓝色	powder blue
浅灰绿色	celandine green
浅灰色	pale grey, light grey, French grey, grayish
浅蓝	Cambridge blue, azury, saxe blue
浅蓝灰色	pearl blue
浅蓝绿色	pale bluish green
浅粒岩	leucoleptite
浅亮黄	light bright yellow
浅绿色	aqua, reseda, jade-green, laurel-green
浅埋成岩作用	shallow burial diagenesis
浅色	pale, tinge, light, light-colored
浅色长英麻粒岩	leucoleptite
浅色矿物	light-colored mineral
浅色体	leucosome
浅色岩	leucocratic rock , leucocrate

浅闪石	edenite
浅滩	ripple
浅锈黄	rust yellow
浅源地震	shallow-focus-earthquake
浅紫色	lilac
欠发达	under-development
欠发达造成的资源短缺	resource scarcity due to under-development
茜素红	alizarin red
嵌晶结构	poikilotopic texture, poikilitic texture
嵌晶状的	poikilitic
嵌状变晶结构	poikiloblastic texture
锖色	tarnish
蔷薇辉石	rhodonite
蔷薇辉石矿床	rhodonite deposit
蔷薇石英	rose quartz
羟钒锌铅石	descloizite
羟磷灰石	hydroxylapatite
羟砷锌石	adamite
羟铁云母	annite
巧克力色	chocolate
鞘褶皱	shearth fold
亲花岗岩矿床	granophile deposit
亲硫元素	sulfophile element, chalcophile element
亲气元素	atmophile element, inert gas
亲生物元素	biophile element
亲石元素	lithophile element
亲铁元素	siderophile element, transitional element
亲铜元素	chalcophile element, sulfophile element
亲氧过程	aerobic process
侵入[作用]	intrusion
侵入接触	intrusive contact
侵入体	intrusive body, intrusion
侵入岩	intrusive rock

侵入岩体	pluton
侵入作用	intrusion
侵蚀[作用]	erosion
侵蚀基面	erosion base level
侵蚀基准面	base level of erosion
侵蚀阶地	erosional terrace
侵蚀作用	erosion
青河石	qingheiite
青金石	lazurite
青铝闪石	crossite
青绿的	blue green
青绿色	turquoise [blue]
青盘岩化	meta-andesitization
青磐岩	propylite
青磐岩化	propylitization
青色	cyan, azury
青石棉矿床	crocidolite deposit, crocidolite-asbestos deposit
青铜黄色的	bronze yellow
青铜蓝	bronze blue
青铜色	bronze, bronzed
青铜烟灰色	bronze smoke
氢	Hydrogen(H)
氢键	hydrogen bond
氢交代作用	hydrogen metasomatism
氢-水体系	H-H_2O system
氢氧化物	hydroxide
氢氧镁石	brucite
氢-氧体系	H-O system
轻敲模式原子力显微镜	Tapping Mode Atomic Force Microscopy (TMAFM)
轻稀土元素	light rare earth element(LREE)
轻质油	light oil

倾伏[角]	plunge
倾伏褶皱	plunging fold
倾角	dip angle
倾竖褶皱	vertical fold
倾向	dip
倾向滑动断层	dip-slip-fault
倾斜层理	inclined bedding
倾斜构造	dipping structure
倾斜型	clinoform
倾斜褶皱	inclined fold
穹隆	dome
丘陵	hill
丘状交错层理	hummocky cross-bedding
球颗	variole
球颗结构	variolite texture, variolitic texture
球颗玄武岩	variolite
球粒	chondrule, pellet
球粒亮晶灰岩	pelsparite
球粒微晶灰岩	pelmicrite
球粒陨石	chondrite
球粒状	spherulitic
球面投影	spherical projection
球面坐标	spherical coordinate
球石	coccolith
球土	ball clay
球形	sphere
球形风化	spheroidal weathering
球状构造	orbicular structure
区域背景	regional background
区域变质(或变质区)*P/T* 比类型	*P/T* ratio types of regional metamorphism (or metamorphic regions)
区域变质[作用]	regional metamorphism

区域变质岩	regional metamorphic rock
区域成矿学	regional metalogeny
区域地球化学测量	regional geochemical survey
区域地球化学异常	regional geochemical anomaly, areal geochemical anomaly
区域动力变质[作用]	regional dynamic metamorphism
区域动力变质岩	regional dynamometamorphic rock
区域动热变质[作用]	regional dynamothermal metamorphism
区域环境	regional environment
区域混合岩化[作用]	regional migmatization
区域静力变质[作用]	regional static metamorphism
曲流河	meandering stream
曲率	curvature
去白云岩化	dedolomitization
去气作用	degasification
全对称性	holohedrism
全晶质	holocrystalline
全面体	holohedron
全面像	holohedrism
全球变暖	global warming
全球海平面	global sea level
全球海平面变化	eustacy
全球环境	global environment
全熔岩	diatexite
全消光	complete extinction
全新世(统)	Holocene Epoch(Series),Holocene
全形	holohedral form
全型	holotype
全岩	bulk rock
全自形粒状	panidiomorphic granular
缺失	lacuna
缺位	vacancy
群	group

群落的季相	seasonal aspect of biome
群落的演退	succession of biome
群青	French blue, ultramarine blue

R

燃料矿产	mineral fuel
燃料矿床	mineral fuel deposit
热[动]力学	thermodynamics
热[致]发光	thermoluminescence
热变质岩	thermometamorphic rock
热变质作用	thermometamorphism
热成熟	thermal ripening
热成熟成因气	gas from thermomaturation of organic matter
热传导率	thermal conductivity
热传导作用	heat transfer process
热带气旋	tropical cyclone
热点	hot spot
热电性	pyroelectricity, thermoelectricity
热动力学	thermodynamics
热峰	thermal peak
热峰等温线	thermal-peak isotherms
热峰等压线	thermal-peak isobars
热峰条件	thermal peak conditions
热峰温度	thermal-peak temperature
热峰压力	thermal-peak pressure
热改造型铀矿	thermal reworked type uranium deposit
热函(焓)	enthalpy (H)
热离子化质谱仪	Thermal Ionization Mass Spectroscopy (TIMS)
热力学的	thermodynamic
热量平衡	heat balance
热流柱	plume

热卤水成矿作用	hot-brine metallogenesis
热卤水库	hot brine reservoir
热膨胀	thermal expansion
热泉	hot spring
热泉矿床	hot spring deposit
热扰动	thermal disturbance
热容	heat capacity
热松弛	thermal relaxation
热压地球化学	thermobarogeochemistry
热液矿床	hydrothermal deposit
热液蚀变	hydrothermal alteration
热液通道	hydrothermal channel
热液作用	hydrothermalism
热重分析	Thermal Gravimetric Analysis (TGA)
热转变	thermal transformation
人-地关系	man-earth relationship
人工矿物	artificial mineral, synthetic mineral
人口爆炸	population explosion
人类起源	origin of humankind
人为排放	anthropogenic discharge
刃脊	knife-edge crest
韧性	tenacity
日本	Japan
日光榴石	helvite
日光石	sunstone
容矿岩	host rock
溶洞	cave
溶洞充填	solution-cavity filling
溶斗	corroded funnel
溶度积	solubility product
溶剂	dissolvent
溶解	solution
溶解[作用]	dissolution

溶解动力学	dissolution kinetics
溶解度	solubility
溶解气	dissolved gas
溶解气驱储油层	depletion drive reservoir
溶蚀	erosion
溶蚀盆地	dissolution basin
溶蚀洼地(溶洼)	solution depression
熔长石	maskelynite
熔结凝灰结构	welded tuff texture
熔结凝灰岩	welded tuff, ignimbrite
熔壳	fusion crust
熔离矿床	liquation deposit
熔炼石英矿床	fused quartz deposit
熔融	melting
熔融曲线	melting curve
熔融石	tectite
熔岩	lava
熔岩被	lava sheet
熔岩地貌	lava landform
熔岩泛流	lava flood
熔岩高原	lava plateau
熔岩管道	lava tube
熔岩湖	lava lake
熔岩流	lava flow
熔岩喷泉	lava fountain
熔岩穹隆	lava dome
熔岩隧道	lava tunnel
熔岩席	lava sheet
揉皱结构	crumpled texture
肉色	incarnadine, incarnate, cuticolor, flesh-colored
肉眼	naked eye
铷	Rubidium(Rb)

铷矿床	rubidium deposit
铷-锶计时	Rb-Sr dating
蠕变性	creep behaviour
蠕虫结构	myrmekitic texture
蠕虫状结构	myrmekitic texture
蠕绿泥石	prochlorite
乳白色	oyster white,ivory
乳蛋白晕彩	milky opalescence
乳滴状结构	emulsion texture
乳光	opalescence
乳金	milk gold
乳石英	milky quartz
软流圈	asthenosphere
软锰矿	pyrolusite
软泥	ooze
软体动物	mollusk
瑞典	Sweden
瑞利分馏	Rayleigh fractionation
瑞士	Switzerland
若特维尔德 X 射线矿物含量分析方法(?)	Rietveld X-ray Modal Analysis Method (RXMAM)

S

SLAP 标准(国际原子能机构推荐的第二个氢、氧同位素标准，δD＝－428‰)	SLAP(standard light Antarctic precipitation)
SMOW 标准(标准平均大洋水，氢、氧同位素标准)	SMOW(standard mean ocean water)
S 面	S-safaces,S-planes
萨勃哈	sabkha
萨布哈成矿模式	sabkha metallogenic model
萨布哈成盐模式	sabkha salt model
萨德伯里铜镍矿床	Sudbury copper-nickel deposit

萨尔瓦多	El Salvador
萨克森蓝	saxe blue
萨斯喀彻温省(加)	Saskatchewan
塞拉里昂	Sierra Leone
塞内加尔	Senegal
塞浦路斯	Cyprus
塞舌尔	Seychelles
三叉点	triple point, triple junction
三叠纪(系)	Triassic Period(System), Triassic
三方单锥	trigonal pyramid
三方晶系	trigonal crystal system, rhombohedral system
三方偏方面体	trigonal trapezohedron
三方双锥	trigonal bipyramid
三方柱	trigonal prism
三级红	3-rd order pink
三级相变	tricritcal order transition
三角三八面体	triakisoctahedron
三角三四面体	triakistetrahedron
三角洲	delta
三联点	triple junction
三水铝石	gibbsite
三维地带性	three dimensional zonality
三维图	three dimentional diagram
三向联结构造	triple point, triple junction
三斜晶系	triclinic system
三斜铁辉石	pyroxferroite
三斜霞石	carnegieite
三元[体]系	ternary system
三元共晶	ternary eutectic
散裂成因核素	spallogenic nuclide
散落物	fallout, airborne debris
扫描电镜	scanning electron microscope (SEM)

扫描电子显微镜	Scanning Electron Microscopy (SEM)
扫描力显微镜	Scanning Force Microscopy (SFM)
扫描隧道光谱	Scanning Tunneling Spectroscopy (STS)
扫描隧道显微镜	Scanning Tunneling Microscopy (STM)
扫描探针显微镜	Scanning Probe Microscopy (SPM)
扫描质子探针分析	Scanning Proton-Microprobe Analysis (SPMA)
色的纯度	saturation
色的亮度	lightness
色率	color index
色散	dispersion
色素离子	chromophores
色心	color centre
铯	Caesium(Cs)
铯矿床	cesium deposit
沙坝	sand bar
沙尘暴	sandstorm
沙堤	sand barrier
沙堆	sand mound
沙漠	sandy desert，deserts
沙漠化	desertification
沙漠区	desert area
沙丘	dunes
沙滩	sand beach
沙特阿拉伯	Saudi Arabia
沙质海岸	sand coast
砂金矿床	gold placer
砂金石(富包体石英)	aventurine
砂金石闪光	aventurescence
砂咀(沙嘴)	sand spit
砂矿床	placer
砂锡矿床	tin placer
砂岩	sandstone

砂岩型铀矿	sandstone type U-ore
砂质板岩	sandy slate
砂钟构造	hourglass texture
筛状结构	sieve texture
山	mountain
山间盆地	intermontane basin
山麓冰川	piedmont glacier
山脉	mountain range
山系	mountain system
山岳冰川	mountain glacier
钐	Samarium(Sm)
钐-钕计时	Sm-Nd dating
珊瑚[虫]	coral
珊瑚礁	coral reef
闪长岩	diorite
闪光	schiller
闪突起	twinkling
闪斜煌岩	spessartite
闪锌矿	sphalerite
闪正煌岩	vogesite
扇积砾	fanglomerate
扇砾岩	fanglomerate
扇形褶皱	fan fold
熵	entropy (S)
上奥陶统	Upper/Late Ordovician
上白垩统	Upper/Late Cretaceous
上半球	upper hemisphere
上寒武统	Upper/Late Cambrian
上泥盆统	Upper/Late Devonian
上盘	hanging wall
上三叠统	Upper/Late Triassic
上升盘	upthrown side
上升溶液	ascending solution

上升洋流成矿作用	upwelling-current mineralization
上升洋流磷矿成矿模式	upwelling-current model of phosphate deposit
上新世(统)	Pliocene Epoch(Series),Pliocene
上侏罗统	Upper/Late Jurassic
烧绿石	pyrochlore
烧绿石碳酸岩	pyrochlore-bearing carbonatite
烧石膏	polyhydrate(bassanite)
蛇绿岩	ophiolite
蛇绿岩套	ophiolite complex, ophiolite suite
蛇丘	esker
蛇纹石	serpentine
蛇纹石-滑石片岩	serpentine-talc schist
蛇纹石化	serpentinization
蛇纹石石棉	serpentine asbestos,chrysotile
蛇纹石石棉矿床	chrysotile-asbestos deposit
蛇纹岩	serpentinite
申(圣)弗利斯符号	Schoenflies symbol
伸展作用	extension
砷	Arsenic(As)
砷矿床	arsenic deposit
砷酸盐	arsenate
砷锑矿	stibarsen
砷锌矿	reinerite
深部构造线成矿说	metallogeny of deep lineaments
深成侵入体	pluton
深成侵入相	plutonic intrusive facies
深成岩	plutonic rock, plutonite
深成岩体	pluton
深带	katazone
深度补偿	depth compensation
深度带	depth zone
深粉红色	radiance

深海	deep-sea
深海的	abyssal, pelagic
深海扇	deep-sea fan
深褐色	puce,dark brown
深黑	jet black,atrous
深红	scarlet,crimson,deep red,carmine, bright red
深黄红色	strong yellowish pink
深灰褐色	dark grayish brown
深蓝色	navy blue,dark blue,blue-black, cyaneous,mazarine,mid night blue
深绿	dark geen,bottle-green
深绿褐色的	olive-drab
深埋成岩作用	deep burial diagenesis
深切河曲	incised meander
深熔[作用]	anatexes, anatexis
深色的	saturate,saturated,dark
深源地震	deep-focus earthquake
深紫红色的	claret-colored
深紫色	modena
深棕色	nut-brown
肾状构造	reniform structure
肾状集合体	reniform aggregate
渗流	seepage flow
渗流区	vadose zone
渗漏晕	leakage halo
渗滤	infiltration
渗滤粉砂	vadose silt
渗透	infiltration
渗透交代[作用]	infiltration metasomatism
渗透率	permeability
渗透性	permeability
渗透混合岩	permeation migmatite

渗透片麻岩	permeation gneiss
升华作用	sublimation
升降运动	vertical movement
生产[油]井	producing well
生产者	producer
生长	growth
生长丘	growth hillocks
生长双晶	growth twin
生长台阶	growth steps
生长条纹	striation(striae)
生成顺序	sequence
生活环境	habitat
生理学	physiology
生命起源	the origine of life
生命元素	bioelement
生态	zoology
生态系统	ecological balance, ecosystem
生态学	ecology
生态因素	ecological factor
生物爆发	biotic explosion
生物标志化合物	biomarker, biological marker compound
生物沉积	biogenic sediment
生物成因气	biogenetic gas
生物成因作用	biogenic process
生物的适应性	biological adaptation
生物的指示现象	biological indicator
生物地层层理	biostratification
生物地层带	biostratigraphic zone
生物地层单位	biostratigraphic unit
生物地层学	biostratigraphy
生物地球化学	biogeochemistry
生物地球化学测量	biogeochemical survey
生物地球化学循环	biogeochemical cycle

生物多样性	bio-diversity
生物风化	biological weathering
生物富集	biological concentration
生物骨架灰岩	framestone
生物化学沉积矿床	mineral deposit by biogenic and biochemical sedimentation
生物化学沉积岩	biochemical sedimentary rock
生物化学风化作用	biochemical weathering
生物降解作用	biodegradation
生物礁	reef, bioherm
生物净化	biological purification
生物矿化作用	biomineralization
生物累积	bioaccumulation
生物亮晶灰岩	biosparite
生物量	bio-mass
生物泥晶灰岩	biomicrite
生物圈	biosphere
生物群落	biome, biotic community
生物扰动构造	bioturbation structure
生物扰(搅)动作用	bioturbation
生物软泥	bioslime
生物碎屑	bioclast
生物微晶灰岩	biomicrite
生物吸收系数	biological absorption coefficient(BAC)
生物系[统]	ecosystem
生物相	biofacies
生物小循环	micro-biological cycle
生物粘结灰岩	boundstone
生物钻孔	burrows
生油层	source bed
生油门限	threshold of oil generation
生油岩	source rock
生油岩评价	source rock evaluation

生油液态窗	liquid window of oil generation
生源物	precursor compound
绳状熔岩	pahoehoe lava，ropy lava
圣多美和普林西比	Sao Tome and Principe
圣赫勒拿岛和阿森松岛等(英)	Saint Helena and Ascension etc.
圣基茨和尼维斯联邦	The Federation of Saint Kitts and Nevis
圣卢西亚	Saint Lucia
圣马力诺	San Marino
圣皮埃尔和密克隆群岛(法)	Saint Pierre and Miquelon
圣文森特和格林纳丁斯	Saint Vincent and The Grenadines
施若内马克思法则	Schreinemakers' rule
施若内马克思束	Schreinemakers' bundle
湿地	wet land，swampy land
湿化学分析法	Wet Chemical Analysis (WCA)
湿气	wet gas，rich gas
十亿分之一(10^{-9})	ppb (part per billion)
十字石	staurolite
石材	dimension stone
石膏	gypsum
石膏化	gypsification
石膏矿床	gypsum deposit
石膏试板	gypsum plate
石化[作用]	lithification
石黄	mineral yellow
石灰华	travertine
石灰石	limestone
石灰岩	limestone
石灰岩矿床	limestone deposit
石蜡	wax，paraffin wax
石榴子石矿床	garnet deposit
石榴子石	garnet
石绿	mineral green
石棉矿床	asbestos deposit

石墨	graphite
石墨矿床	graphite deposit
石泡构造	lithophysa structure
石蕊蓝	litmus blue
石笋	stalagmite
石笋状	stalagmitic
石炭纪(系)	Carboniferous Period(System), Carboniferous
石铁陨石	stony-iron meteorite
石盐	halite, rock salt
石盐矿床	halite deposit
石印蓝(无游离硫的群青，适用于油墨)	lithosized blue
石英	quartz
石英安山岩	quartz andesite
石英斑岩	quartz porphyry
石英粗安岩	quartz trachyandesite
石英粗面岩	quartz trachyte
石英二长岩	quartz monzonite
石英化	quartzification
石英角斑岩	quartz-keratophyre
石英净砂岩	quartz arenite
石英喷流岩	quartz exhalite
石英砂矿床	quartz sand deposit
石英砂岩	quartz sandstone
石英闪长岩	quartz diorite
石英楔	quartz wedge
石英岩	quartzite
石英岩砾岩	quartzitic conglomerate
石油产量	oil production rate
石油产状	oil occurrence
石油初次运移	oil primary migration
石油地球化学	petroleum geochemistry

石油地质学	petroleum geology
石油二次运移	oil secondary migration
石油聚集	oil accumulation
石陨石	stony meteorite
石钟乳	stalactite
石柱	stalacto-stalagmite
时	chron
时带	chronozone
时间温度指数	time temperature index(TTI)
实际矿物成分	mode
实验地球化学	experimental geochemistry
实验矿床学	experimental metallogeny
实验矿物学	experimental mineralogy
实验室分析	laboratory analysis
实验岩石学	experimental petrology
蚀变[作用]	alteration
蚀变花岗岩铌钽矿床	niobium and tantalum deposit in altered granite
蚀变围岩	altered wallrock
蚀变岩[石]	altered rock
蚀坑	etch pit
蚀象	etch figure
蚀源区	provenance
食物	food web
食物链	food chain
矢量	vector
始太古界	Eoarchean
始新世(统)	Eocene Epoch(Series),Eocene
世	epoch
世代	generation
视地热梯度	apparent geothermal gradient
视倾角	apparent dip
视倾斜	apparent dip

试板	test plate
试点测量	orientation survey
试生产测量	pilot survey
饰边	fringe
铈	Cerium(Ce)
铈异常	Ce anomaly
收缩构造	contraction tectonics
收缩说	contraction theory
手标本	hand specimen
寿山石	soapstone
受变质沉积铁矿床	metamorphosed sedimentary iron deposit
受变质硅铁建造	metamorphosed cherty iron formation
受变质矿床	metamorphosed deposit
瘦煤	lean coal
枢纽	hinge
枢纽带	hinge zone
枢纽线	hinge line
梳状构造	comb structure
束状集合体	packet aggregate
束状结构	sheaf texture
树枝状集合体	dendritic aggregate
树脂光泽	resinous luster
树脂体	resinite
数字地球	digital earth
双变平衡	divariant equilibrium
双变区	divariant field
双变质带	paired metamorphic belts
双岛状硅酸盐	sorosilicate
双断裂夹持区	sandwiched area of double fracture zone
双反射	bireflection
双硅酸盐	disilicate
双交代[作用]	bimetasomatism
双晶	twin crystal, twinning

双晶缝合线	twin partition line
双晶接合面	composition plane, composition surface
双晶律	twin law
双晶面	twinning plane
双晶片	twin lamellae
双晶条纹	twinning striation
双晶要素	twin element
双晶中心	twin center
双晶轴	twin axis(twinning axis)
双链状硅酸盐	double chain silicate
双目显微镜	Stereoscopic Microscopy (SM)
双圈测角仪	two circle goniometer
双向的转变	enantiotropic transformation
双折射率	birefringence
双折射率色散	dispersion of birefringence
双锥	bipyramid, dipyramid
霜	frost
水成碎屑岩	hydroclastic rock
水地球化学测量	geochemical water survey
水动力	hydrodynamic force
水和二氧化碳参加的反应	reactions that involve H_2O and/or CO_2
水化[作用]	hydration
水化的	hydrous
水解	hydrolysis
水解的	hydrolysed
水晶	rock crystal
水库	reservoir
水蓝	saxe blue
水流蚀痕	tool marks
水镁石	brucite
水锰矿	manganite
水泥石灰岩	cement limestone
水泥原料	cement raw materials

水泥原料矿产	raw material for cement
水平层理	horizontal bedding
水平分带	horizontal zoning
水平构造	horizontal structure
水平双晶	acline twins
水平运动	horizontal movement
水平褶皱	horizontal fold
水驱储油层	water drive reservoir
水圈	hydrosphere
水热对流单元	convective hydrothermal cell
水热反应	hydrothermal reaction
水热矿床	hydrothermal deposit
水热流体	hydrothermal fluid
水热喷发	hydrothermal eruption
水热蚀变	hydrothermal alteration
水热通道	hydrothermal channel
水热系统	hydrothermal system
水热钻石高压盒	Hydrothermal Diamond-Anvil Cell (HDAC)
水碳硼石	carboborite
水土病	acclimation disease
水土流失	soil erosion
水文地球化学	hydrogeochemistry
水文循环	hydrological cycle
水污染	water pollution
水系沉积物测量	stream sediment survey
水系异常	drainage anomaly
水下的	subaqueous
水下滑移[作用]	subaqueous gliding(subaqueous slump)
水星叶石	hydroastrophyllite
水岩作用	water-rock interaction
水俣病	Minamata disease
水云母	hydromica

水云母化	hydromicazation
顺层混合岩	epibolite
顺磁性	paramagnetism
顺流	downcurrent
顺向分带	normal zoning
顺序提取	sequential extraction
瞬时等温线	instantaneous isotherms
瞬时等压线	instantaneous isobars
瞬时地热梯度	transient geothermal gradient
丝带结构	ribben texture
丝光沸石	mordenite
丝绢光泽	silky luster
丝炭	fusain
丝炭化作用	fusinization
丝质体	fusinite
斯里兰卡	Sri Lanka
斯洛伐克	Slovakia
斯洛文尼亚	Slovenia
斯石英	stipoverite
斯威士兰	Swaziland
锶	Strontium(Sr)
锶富集指数	delta Sr
锶-氧体系	Sr-O system
死火山	extinct volcano
四方单锥	tetragonal pyramid
四方晶系	tetragonal system
四方偏方面体	tetragonal trapezohedron
四方双锥	tetragonal bipyramid
四方双锥(楔)	tetragonal disphenoid
四方四面体	tetragonal tetrahedron
四方铜金矿	tetraauricupride
四方柱	tetragonal prism

四角三八面体	tetragonal trisoctahedron, tetrakisoctahedron, trapezohedron
四角三四面体	tetrakistetrahedron
四六面体	tetrahexahedron, tetrakishexahedron
四面体	tetrahedron
四元[体]系	quaternarysystem
似层状矿脉	bedded wein
似长石	feldspathoids
似核形石	oncoid
似辉石	pyroxenoid
似团粒	peloid
松山极性时	Matuyama polarity chron
松脂光泽	resinous luster
松脂岩	pitchstone
苏必利尔湖型铁矿床	Lake Superior-type iron deposit
苏长岩	norite
苏打湖	natron lake
苏丹	Sudan
苏里南	Suriname
苏斯效应	Suess effect
塑性	plasticity
塑性变形	plastic deformation
塑性流动	plastic flow
塑性行为	plastic behavior
溯源侵蚀	headward erosion
酸碱度	PH(acidity-alkalinity)
酸性	acidic, silicic
酸性白土	acid clay
酸性的	acidic, silicic
酸性环境	acid environment
酸性岩	acidic rock
酸雨	acid precipitation, acid rain
随机采样	random sampling

岁差	precession
碎斑	porphyroclast
碎玻质熔岩	hyaloclastite
碎裂结构	cataclastic texture
碎裂流动	cataclastic flow
碎裂岩	cataclasite
碎石[堆]	debris
碎屑	fragment,clast,debris
碎屑的	detrital
碎屑惰质体	inertodetrinite
碎屑腐植体	humodetrinite
碎屑结构	detrital texture
碎屑镜质体	vitrodetrinite
碎屑壳质体	liptodetrinite
碎屑流	debris flow
碎屑稳定体	liptodetrinite
燧石	flint,chert
燧石铁建造	cherty iron formation
燧石状	flinty
缩小	reduce
所罗门群岛	Solomon Islands
索伦石	suolunite
索马里	Somalia

T

他色	allochromatic color
他生沉积岩	allogenic (sedimentary) rock
他生矿物	allogenic mineral
他形粒状	xenomorphic granular, allotriomorphic granular
他形粒状结构	allotriomorphic granular texture
他型	allotype
它形变晶的	allotrioblastic

铊	Thallium(Tl)
铊矿床	thallium deposit
塔吉克斯坦	Tajikistan
台风	typhoon
苔绿色	moss green
苔藓虫[类]	bryozoan
苔癣	lichen
太古代陆核	Archean nucleus
太古宙(宇)	Archaean Eon(Eonothem),Archaean
太阳常数	solar constant
太阳风	solar wind
太阳系	solar system, sun system
太阳系元素丰度	solar system abundance of element
钛	Titanium(Ti)
钛尖晶石	ulvospinel
钛矿床	titanium deposit
钛铁矿	ilmenite
泰国	Thailand
酞菁蓝	Heliogen blue
滩岩	beachrock
弹性	elasticity
弹性行为	elastic behavior
坦桑尼亚	Tanzania
钽	Tantalum(Ta)
炭灰色	charcoal gray
探井	exploratory well
探明储量	proven reserve
探途元素	pathfinder element
碳	Carbon(C)
碳硅泥岩型铀矿	carbonaceous siliceous-pelitic rock type U-ore
碳磷灰石	carbonate-apatite
碳纳米管	carbon nanotube

碳氢化合物	hydrocarbon
碳酸化	carbonation
碳酸岩	carbonatite
碳酸岩型稀土矿床	carbonatite-type rare earth deposit
碳酸盐	carbonate
碳酸盐补偿深度	carbonate compensation depth(CCD)
碳酸盐沉淀动力学	carbonate precipitation kinetics
碳酸盐台地	carbonate platform
碳酸盐岩	carbonate rocks
碳酸盐岩层控铅锌矿床	stratabound lead-zinc deposit in carbonate rocks
碳酸盐岩化	carbonatization
碳酸盐岩隆	carbonate buildup
碳酸盐岩铅锌交代矿床	metasomatic lead-zinc deposit in carbonate rock
碳循环	carbon cycle
碳质球粒陨石	carbonaceous chondrite
汤加	Tonga
唐奈-哈克原理	Danny-Harker rule
糖粒状白云岩	sucrosic dolomite
糖粒状大理岩	saccharoidal marble
桃[红]色	peach,peachblow,peach-colored,peachy
陶瓷原料	ceramic raw materials
陶瓷原料矿产	ceramic material
陶粒原料	haydite materials
特堤斯成矿带	Tethyan metallogenic belt
特克斯和凯科斯群岛(英)	Turks and Caicos Islands
特拉华州(美)	Delaware
特立尼达和多巴哥	Trinidad and Tobago
特殊[单]形	special form
特殊重力测试	Specific Gravity Determinations (SG)
特提斯[海]	Tethys
特征矢量	eigenvector

铽	Terbium(Tb)
腾冲铀矿	tengchongite
藤黄(浅亮黄)	rattan yellow(a light bright yellow)
梯状脉	ladder vein
锑	Antimony(Sb)
锑矿床	antimoney deposit
体波	body wave
体积系数(原油体积系数)	formation volume factor
体心格子	body-centered lattice, I-lattice
天河石	amazonite
天河石化	amazonitization
天蓝	cerulean,azure,celeste,ciel,sapphire, sky blue,celeste,skyey
天蓝石	lazulite
天气	weather
天青石	celestite
天球	celestial sphere
天然碱	trona
天然碱矿床	trona deposit
天然焦	natural coke
天然气	natural gas
天然气地球化学	natural gas geochemistry
天然气液	natural gas liquid
天体	celestial body
天体地质学	astrogeology
天体化学	astrochemistry
天体演化理论	cosmogonic theory
天文单位	astronomical unit(AU)
田间持水量	field capacity
田纳西州(美)	Tennessee
填积[作用]	aggradation
填间结构	intersertal texture
填隙物质	matrix

条带状赤铁石英岩	banded hematite quartzite
条带状磁铁石英岩	striped magnetite quartzite
条带状大理岩	banded marble
条带状分异岩	banded differentiate
条带状构造	banded structure, banding
条带状硅铁建造	banded cherty iron formation
条带状混合岩	striped migmatite
条带状石英岩	banded quartzite
条痕	streak
条痕状混合岩	streaky migmatite
条痕状片麻岩	streaky gneiss
条件储量	conditional reserve
条纹	fringe, lamella
条纹层理	flaser bedding
条纹长石	perthite
条纹结构	perthitic texture
调制结构	modulated structure
跳跃	saltation
跳跃搬运负荷	saltation load
铁	Iron(Fe)
铁白云石	ankerite
铁磁性	ferromagnetism
铁橄榄石	fayalite
铁硅白云母	ferriphengite
铁滑石	minnesotaite
铁辉石	ferrosilite
铁尖晶石	hercynite
铁矿床	iron deposit
铁锂云母	zinnwaldite
铁菱镁矿	breunerite
铁铝石榴子石	almandite
铁帽	gossan
铁帽型铁矿床	gossan-type iron deposit

铁镁矿物	mafic mineral
铁镁指数	ferromagnesian index
铁闪石	grunerite
铁蛇纹石	greenolith
铁纹石	kamacite
铁锈色的	ferrugineous, rusty
铁岩	ironstone
铁叶云母	siderophyllite
铁英岩	itabirite, taconite
铁陨石	iron meteorite
铁质石英岩	ferruginous quartzite
烃	hydrocarbon
烃源层	source bed
烃源岩	source rock
同步辐射 X 射线形貌分析	Synchrotron X-ray Topography (SXRT)
同步辐射 X 射线荧光探针分析	Synchrotron X-ray Fluorescence Microprobe Analysis (SXFM)
同构造变晶	syntectonic blast
同化[作用]	assimilation, anabolism
同化混染与分离结晶[作用]	assimilation and fractional crystallization (AFC)
同极	homopolar
同结构	isostructure
同结晶	cotectic crystallization
同生沉积的	syndepositional
同生成矿说	syngenetic ore-forming theory
同生成矿作用	syngenetic ore-forming process
同生矿床	syngenetic deposit
同生水热作用	syngenetic hydrothermal process
同生异常	syngenetic anomaly
同位素地球化学	isotope geochemistry
同位素地温计	isotope geothermometer
同位素地质年代学	isotopic geochronology

同位素分馏	isotopic fractionation
同位素分馏系数	isotope fractionation factor
同位素交换反应	isotopic exchange reaction
同位素交换平衡	isotopic exchange equilibrium
同位素平衡	isotopic equilibrium
同位素稀释方法	isotope dilution method
同位素稀释热离子化质谱仪	Isotope Dilution Thermal Ionization Mass Spectroscopy (IDTIMS)
同斜[层]	homocline
同心鲕	concentric ooid
同心环状构造	concentric structure
同型	homotype
同质多象	polymorphism
同质多象变体	polymorph
同质多象转变	polymorphic transformation
同质二象	demorphism
同质假象	paramorphism
同质三象	trimorphism
桐柏矿	tongbaite
铜	Copper(Cu)
铜黄色	brassy yellow
铜矿床	copper deposit
铜蓝	covellite
铜镍矿床	copper-nickel deposit
铜镍硫化物矿床	copper-nickel sulphide deposit
铜砷铀云母	zeunerite
铜铀云母	torbernite
统	series
统计岩石学	statistical petrology
痛痛病	itai-itai disease
透长石	sanidine
透长岩	sanidinite
透长岩相	sanidinite facies

透辉石	diopside
透镜状	lenslike
透镜状层理	lenticular bedding
透镜状矿脉	lensoid ore body
透锂长石	petalite
透明	transparent
透明度	transparency
透明矿物	transparent mineral
透明性	diaphaneity
透明纸	cellophane paper
透闪石	tremolite
透射电子显微镜	Transmission Electron Microscopy (TEM)
透水层	permeable bed
透水性	water permeability
突尼斯	Tunisia
突起	relief
图幅	sheet
图解表示	graphic representation
图解法	graphic solution
图例	legend
图瓦卢	Tuvalu
图像	picture
土地沙漠化	land desertization
土耳其	Turkey
土耳其玉色	turquoise blue
土黄	stone yellow
土库曼斯坦	Turkmenistan
土壤	soil
土壤地球化学测量	goechemical soil survey
土壤肥力	soil fertility
土壤分类	soil classification
土壤胶体	soil colloid
土壤孔隙度	soil porosity

土壤剖面	soil profile
土壤侵蚀	soil erosion
土壤圈	pedosphere
土壤异常	soil anomaly
土体	solum
土状光泽	earthy luster
土状集合体	earthy aggregate
钍	Thorium(Th)
钍矿床	thorium deposit
钍铀比	Th/U ratio
湍流	turbulent flow
团块	clot
团粒	pellet
推覆构造	nappe structure
推覆体	nappe
推移[作用]	traction
退白云岩化	dedolomitization
退变岩	diaphthorite
退变质	retrograde metamorphism
退变质岩	retrograde metamorphic rock
退化变质	retrogressive metamorphism
退流	abstraction
蜕变	disintegration
蜕晶作用	metamictization
褪色作用	decolorization
托克劳群岛(新)	Tokelau Islands
脱玻化[作用]	devitrification
脱挥发分反应	devolatilization reaction
脱水	dewatering
脱水反应	dehydration reaction
脱水反应等变线	dehydration-reaction isograd
脱水实验	dehydration experiment
脱水作用	dehydration

脱碳酸反应	decarbonation reaction
驼色	camel
椭圆偏振	elliptical polarization
拓扑学	topology
拓扑学的	topological

U

U 形谷	U-shaped valley

V

VSMOW 标准(维也纳标准平均大洋水,国际原子能机构推荐的氢、氧同位素标准,δD=0)	VSMOW (Vienna standard mean ocaen water)

W

挖掘作用	glacial sapping
瓦板岩	roofing slate
瓦克岩	wacke
瓦利斯群岛和富图纳群岛(法)	Wallis and Futuna Islands
瓦努阿图	Vanuatu
歪长石	Anorthoclase
歪晶	distorted crystal
外[来]碎屑	epiclast,extraclast
外动力	exogenetic force
外核	outer core
外缓冲	external buffering
外缓冲组分	externally buffered components
外来[岩]体	allochthon
外生成矿作用	exogenenic metallization
外生矿床	exogenic mineral deposit
外生作用	exogenic process,exogenesis
外延的	epitaxic,epitaxial
外延生长	epitaxial growth,overgrowth
外营力	exogenetic force

外源的	allogenetic
弯滑褶皱	flexural-slip fold
弯流褶皱	flexural-flow fold
弯褶皱	buckle fold
湾流	gulf steam
完全活动组分	perfectly mobile components
完全解理	good cleavage
完全类质同象	complete isomorphous series
顽辉石球粒陨石	enstatite chondrite
顽火辉石	enstatite
烷基苯系列化合物	alkyl benzenes
烷基菲系列化合物	alkyl phenanthrenes
烷基联苯化合物	alkyl biphenyls
烷基萘系列化合物	alkyl naphthalenes
烷烃馏分	aliphatic fraction
晚期残余岩浆型矿床	late residual magma-type deposit
晚期成岩[作用]	late diagenesis
晚期岩浆矿床	late magmatic deposit
万亿分之一(10^{-12})	ppt(part per trillion)
腕足动物门	brachiopoda
网格构造	boxwork
网脉	stockwork,network deposit,stringer lode
网脉状矿石带	stockwork ore zone
网状河流	braided stream
危地马拉	Guatemala
威尔逊旋回	Wilson cycle
威尼斯蓝	Venetian blue,Italian blue
威斯康星州(美)	Wisconsin
威特沃特斯兰金-铀砾岩	Witwatersrand auri-uraniferous conglomerate
微[带]红[色]的	reddish
微暗煤	durite
微玻璃陨石	microtektite

微惰煤	inertite
微分干涉相衬显微镜	Differential Interference Contrast Microscopy (DIC)
微分激光干涉仪	Differential Laser Interferometry (DLI)
微迹元素	trace element
微晶	microlite
微晶石英型铀矿	U-ore of microcrystalline quartz type
微镜惰煤	vitrinertite
微镜煤	vitrite
微壳煤	liptite
微粒的	micrograined
微粒体	micrinite
微亮煤	clarite
微量元素	trace element, microelement
微劈石	microlithon
微嵌晶状的	micropoikilitic
微三合煤	trimacerite
微生物	microorganism
微生物地质过程	microbio-geological process
微束拉曼光谱仪	Microbeam Raman Spectrometry(MBRS)
微稳定煤	liptite
微斜长石	microcline
微斜长石化	microclinization
微型浮游生物	nannoplankton
微异地生成煤	hypautochthonous coal
围山矿	weishanite
围岩	country rock, wall rock, inclosing rock
围岩蚀变	country rock alteration, wallrock alteration
唯象学	phenomenology
维德曼施泰滕相	Widmannstatten pattern
维克斯显微硬度	Vicker's microhardness
维氏显微硬度	Vicker's microhardness
维氏硬度计	Vickers Hardness Number(VHN))

伟晶的	pegmatitic
伟晶岩	pegmatite
伟晶岩矿床	pegmatite mineral deposit
伟晶岩锂矿床(锂辉石-锂云母伟晶岩矿床)	spodumene-lepidolite-pegmatite deposit
伟晶岩型铀矿	pegmatite uranium deposit
伟晶作用	pegmatitization
纬度效应	latitude effect
委内瑞拉	Venezuela
萎蔫点	wilting point
卫星	satellite
未固结的	unconsolidated
位错	dislocation
位移	displacement
位移方向	sense of displacement
位移式相变	displacive transformation
位移型转变	displacive transformation
魏尔纳斯基	Vernasky
温差风化	thermal weathering
温度梯度	temperature gradient
温泉	hot spring
温石棉	chrysotile
温室气体	greenhouse gas
温室效应	greenhouse effect
温压地球化学	thermobarogeochemistry
瘟疫	pestilence
文莱	Brunei
文洛克统	Wenlock
文石	aragonite
文象结构	graphic texture
纹层(理)	lamina，lamella,lamination
纹理年代学	varve chronology
纹泥	varved clay，varve

紊流	turbulent flow
稳[定状]态	stable state
稳定地块	craton
稳定陆块内的	intracratonic
稳定同位素比率质谱仪	Stable Isotope Ratio Mass Spectrometry (SIRMS)
稳定同位素地球化学	stable isotope geochemistry
稳定性	stability
稳定组	exinite
稳态地热梯度	steady-state geothermal gradient
瓮蓝	vat blue
沃克环流	Walker cell
乌干达	Uganda
乌克兰	Ukraine
乌拉尔统	Cisuralian
乌拉圭	Uruguay
乌兹别克斯坦	Uzbekistan
污染	pollution
污染剂量	pollution dose
污染类型	pollution type
污染源	pollution sources
污染指数	pollution index
钨	Tungsten(W)
钨矿床	tungsten deposit
钨锰铁矿	wolframite
钨酸盐	tungstate
钨锑金石英脉	tungsten-antimony-gold-quartz vein
无定向构造	nondirectional structure
无定形的	amorphous
无骨骼(架)的	non-skeletal
无骸晶的	non-skeletal
无机成因气	inorganic genetic gas
无机地球化学	inorganic geochemistry

无机起源	inorganic origin
无结构腐植体	humocollinite
无结构镜质体	collinite
无结构铁陨石	ataxite
无球粒陨石	achondrite
无色的	colorless
无水石膏	anhydrite
无序结构	disorder structure
无烟煤	anthracite
无意义异常	non-significant anomaly
吴氏网	Wulff net
五角三八面体	pentagonal icositetrahedron, pentagonal trisoctahedron
五角三四面体	pentagonal tristetrahedron
五角十二面体	pentagonal dodecahedron, pyritohedron
五元素建造矿床(镍-钴-银-铋-铀建造矿床)	ore deposit of penta-element formation (ore deposit of Ni-Co-Ag-Bi-Uformation)
物理地球化学	physical geochemistry
物理风化	physical weathering
物理模拟	physical model
物质流	material flow
物种灭绝	species extinct
雾	fog
雾灰色	misty gray

X

X射线单晶衍射	X-ray Single Crystal Diffraction (XRSCD)
X射线发光	X-ray luminescence
X射线发光分析	X-ray Luminescence (XRL)
X射线粉晶衍射	X-ray Powder Diffraction (XRPD)
X射线光电子光谱	X-ray Photoelectron Spectroscopy (XPS)
X射线光谱	X-ray Spectrometry (XRS)

X射线激光磨蚀电感偶合质谱仪	X-ray Laser Ablation Inductively Coupled Mass Spectroscopy (XLA-ICP-MS)
X射线晶体学	X-ray crystallography
X射线吸收光谱	X-ray Absorption Spectroscopy (XRAS)
X射线吸收近边结构分析	X-ray Absorption Near-Edge Structure Analysis (XANES)
X射线吸收精细结构光谱	X-ray Absorption fine-Structure Spectroscopy (XAFS)
X射线衍射	X-ray Diffraction (XRD)
X射线荧光谱	X-ray Fluorescence Spectroscopy (XRFS)
矽卡岩	skarn
矽卡岩白钨矿矿床	scheelite skarn deposit
矽卡岩矿床	skarn mineral deposit
矽卡岩硼矿床	boron-bearing skarn deposit
矽卡岩型钼矿床	skarn molybdenum deposit
矽卡岩型铅锌矿床	lead-zinc deposit in skarn
矽卡岩型铁矿床	skarn iron deposit
矽卡岩型铜矿床	skarn copper deposit
夕线石	sillimanite
夕线石片麻岩	sillimanite gneiss
西班牙	Spain
西北地区(加)	Northwest Territories
西弗吉尼亚州(美)	West Virginia
西里西亚式铅锌矿床	Silesian-type lead-zinc deposit
西盟石	ximengite
西撒哈拉	Western Sahara
西萨摩亚	Western Samoa
吸附能	attachment kinetics
吸附水	hydroscopic water
吸附铀	adsorption uranium
吸附作用	adsorption
吸收光谱	Absorption Spectroscopy (AS)
吸收能力	absorptive capacity

吸收性	absorption, absorptivity
吸收指数	absorption index
希尔特规律	Hilt's rule
希腊	Greece
硒	Selenium(Se)
硒矿床	selenium deposit
稀释剂	spike
稀土-磁铁矿-赤铁矿矿床	rare-earth-bearing magnetite-hematite deposit
稀土金属矿床	rare earth metal deposit
稀土碳酸岩矿床	rare-earth-bearing carbonatite deposit
稀土元素	rare earth element(REE)
稀土元素分布模式	distribution pattern of rare earth element
稀土元素矿床	ore deposit of rare-earth elements
稀有碱金属	rare alkaline metal
稀有元素	rare element
稀有元素矿床	ore deposit of rare elements
锡	Tin(Sn)
锡金	Sikkim
锡矿床	tin deposit
锡林郭勒矿	xilingolite
锡石	cassiterite
锡石-硫化物矿床	cassiterite-sulphide deposit
锡石-石英脉	cassiterite-quartz vein
锡铁山石	xitieshanite
膝折	kink
膝折褶皱	kink fold
膝状双晶	knee shaped twin, elbow twin
席状矿体	manto
喜峰矿	xifengite
喜马拉雅期	Himalayan
系	System
系数	coefficient

系统组分	system components
细碧岩	spilite
细槽	groove
细层	lamina
细晶结构	aplitic texture
细晶石	microlite
细晶岩	aplite
细菌作用	bacteria action
细砾岩	granule conglomerate
细粒	fine-grained
细粒长石石英净砂岩	fine feldspathic quartzarenite
细粒长英粒岩	halle-flinta
细粒的	fine-grained, fine grained
细粒化[作用]	fining
细粒结构	fine grained texture
细脉浸染型钼矿床	stockwork and disseminated molybdenum deposit
细脉浸染型铜矿床	veinlet-disseminated copper deposit
细砂岩	fine sandstone
狭带系	Stenian
霞石	nepheline
霞石化	nephelinization
霞石岩	Nephelinite
霞石正长岩	nepheline syenite
霞斜岩	Theralite
下奥陶统	Lower /Early Ordovician
下白垩统	Lower /Early Cretaceous
下半球	lower hemisphere
下垫面	underlying surface
下寒武统	Lower /Early Cambrian
下降盘	downthrown side
下降溶液	descending solution
下泥盆统	Lower /Early Devonian

下盘	footwall
下切[作用]	down-cutting
下三叠统	Lower /Early Triassic
下侏罗统	Lower /Early Jurassic
夏威夷州(美)	Hawaii
先成河	antecedent river
纤钡锂石	balipholite
纤锰柱石	carpholite
纤闪石化	uralitization
纤蛇纹石	chrysotile
纤维状断口	fibrous fracture
纤维状集合体	fibrous aggregate
纤状变晶结构	nematoblastic texture
氙	Xenon(Xe)
鲜粉红色	shocking pink
鲜红	bright red,scarlet,crimson
鲜绿色	emerald green
鲜艳的	colorful
显晶集合体	phanerocrystalline aggregate
显生宙(宇)	Phanerozoic Eon(Eonothem),Phanerozoic (PH)
显微构造	microstructure
显微光度计	microphotometer
显微光子诱导 X 射线发射光谱	Micro Proton-Induced X-ray Emission Spectroscopy (M-PIXE)
显微拉曼光谱	Micro-Raman Spectroscopy (MRS)
显微煤岩类型	microlithotype
显微热分析	Microthermometric Analysis (MTA)
显微探针分析	Microprobe Analysis (MPA)
显微条纹长石(光学显微镜下)	microperthite
显微文像结构的	micrographic
显微压痕硬度测试	Microindentation Hardness Testing (MIHT)

显微照相的	micrographic
显微组分	maceral
现代构造运动	modern tectonic movement
现实主义[原则和方法]	actualism
现象学	phenomenology
线金属量	linear productivity
线理	lineation
线性构造	lineament, linear struscure
线状形迹	lineament
线状褶皱	linear fold
陷落地震	collapse earthquake
霰石	aragonite
相差显微镜	Phase Contrast Microscopy (PCM)
相对丰度	relative abundance
相对密度	relative density
相对渗透率	relative permeability
相对吸收系数	relative absorption coefficient(RAC)
相容元素	compatible element
相似褶皱	similar fold
相序	facies sequences
相组分	phase components
香花石	hsianghualite
湘江铀矿	xiangjiangite
箱状构造物	boxwork
箱状褶皱	box fold
镶嵌结构	mosaic texture
镶嵌式胶结	mosaic cementation
详探井	evaluation well, appraisal well
响岩	phonolite
响岩结构	phonolitic texture
向斜	syncline
向形	synform
象牙[白]色	ivory

消费者	consumer
消光	extinction
消光角	extinction angle
消光位	extinction direction, direction of extinction
消减带	subduction zone
消色器	compensator
硝化作用	nitrification
硝酸盐	nitrate
小麦[黄]色	wheat
小行星	asteroid
肖钠长石	pericline
肖钠长石律双晶	pericline law
楔形	sphenoid
楔状交错层理	wedge cross-bedding
斜长花岗岩	plagiogranite
斜长片麻岩	plagiogneiss
斜长石	plagioclase
斜长岩	anorthosite, plagioclasite
斜发沸石	clinoptilolite
斜方单锥	orthorhombic pyramid, rhombic pyramid
斜方橄榄岩	harzburgite
斜方硅钙石	kilchoanite
斜方辉石	orthopyroxene
斜方辉石岩	orthopyroxenite
斜方晶系	orthorhombic system
斜方闪叶石	ortholamprophyllite
斜方双锥	orthorhombic bipyramid, rhombic bipyramid
斜方四面体	orthorhombic disphenoid, rhombic disphenoid
斜方柱	orthorhombic prism, rhombic prism
斜锆石	baddeleyite
斜硅钙石	larnite
斜绿泥石	clinochlore

斜坡沉积	clinoform
斜坡地形	clinoform
斜歪褶皱	inclined fold
斜顽辉石	clinoenstatite
斜卧褶皱	reclined fold
斜向滑动断层	oblique-slip-fault
斜消光	oblique extinction
斜黝帘石	clinozoisite
鞋带状砂岩	shoestring sandstone
泻湖	lagoon
泻利盐	epsomite
榍石	sphene，titanite
心射极平投影	gnomonic projection
心滩	channel bar
锌	Zinc(Zn)
锌赤铁矾	zincobotryogen
锌氯钾铁矾	zincovoltaite
锌叶绿矾	zincocopiapite
新不伦瑞克省(加)	New Brunswick
新陈代谢	metabolism
新陈代谢作用	metabolism
新成结晶	neocrystallization
新成体	neosome
新第三纪(系)	Neogene Period(System)
新构造运动	neotectonic movement
新罕布什尔州(美)	New Hampshire
新加坡	Singapore
新近系	Neogene
新喀里多尼亚(法)	New Caledonia
新墨西哥州(美)	New Mexico
新生[变]形的	neomorphic
新生变晶	neoblast
新生代(界)	Cenozoic Era(Erathem),Cenozoic (CZ)

新生矿床	juvenile mineral deposit
新斯科舍省(加)	Nova Scotia
新西兰	New Zealand
新仙女木事件	Younger Dryas event
新元古Ⅲ系	Neoproterozoic Ⅲ
新元古代(界)	Neoproterozoic Era(Erathem), Neoproterozoic
新月形沙丘	barchan
新泽西州(美)	New Jersey
信风	trade wind
兴安石	xinganite, hingganite
星彩	asterism
星系	galaxy
星系团	cluster of galaxies
星云	nebula
星云说	nebular hypothesis
星云状构造	nebulitic structure
星状沙丘	star dune
猩红	bright red, scarlet
行迹	trails
行星地质学	planetary
行星际尘(埃)	interplanetary dust
行星际物质	interplanetary matter
形变	distortion
形成年龄	formation age
形貌	morphology
形态	morphology
形态优选方位	dimensional preferred orientation
杏仁构造	amygdaloidal structure
杏仁体	amygdaloid
匈牙利	Hungary
匈牙利蓝	Hungary blue
雄黄	realgar

雄黄矿床	realgar deposit
休顿洛契连生(斜长石出溶形成)	Huttenlocher intergrowth
休眠火山	dormant volcano
溴	Bromine(Br)
叙利亚	Syria
絮凝	flocculation
玄武安山岩	basaltic andesite
玄武球颗结构	variolite texture
玄武碎屑岩	hyaloclastite
玄武土	wacke
玄武岩	basalt
玄武岩风化物	wacke
玄武质	basaltic
玄武质砂岩	wacke
悬臂	spiral arm
悬浮	suspend
悬浮负载	suspended load
悬浮粒	suspended particle
悬浮总体	suspended load
悬谷	hanging valley
悬挂式胶结	pendant cementation
悬移[作用]	suspension transport
悬运	suspension
旋回	cycles
旋绕层理	convolute bedding
旋转法	rotation method
旋转反伸	roto-inversion
旋转反映	roto-reflection
旋转极	pole of rotation
旋转椭球体	rotational ellipsoid
旋转针台	spindle stage
旋转轴	rotation axis
漩涡星系	spiral galaxy

选区电子衍射	Selected-Area Electron Diffraction (SAED)
穴斑沉积物	burrow mottled sediment
雪	snow
雪白色	snowy white
雪球结构	snowball texture
雪线	snow line
循环反应	cyclic reaction
汛期	flood period

Y

压扁[作用]	flattening
压扁层理	flaser bedding
压电石英矿床	piezoquartz deposit
压电性	piezoelectricity
压电原料矿产	raw material for piezoelectricity
压力仓	pressure vessel
压力裾	pressure fringe
压力梯度	pressure gradient
压力校正	pressure calibration
压力影	pressure shadow
压溶	pressure solution
压实(紧)	compaction
压碎角砾岩	crush breccia
压碎结构	crush texture
压碎砾岩	crush conglomerate
压缩	compaction
牙买加	Jamaica
哑层	barren bed
亚结晶的(结晶程度不好的)	paracrystalline(poorly crystalline)
亚颗粒	subgrain
亚拉巴马州(美)	Alabama
亚利桑那州(美)	Arizona

亚麻[黄]色	flaxen
亚美尼亚	Armenia
亚铁磁性	ferrimagnetism
亚铁钠闪石	arfvedsonite
亚稳的	metastable
亚种(矿物分类单位)	subspecies
氩	Argon(Ar)
氩-氩计时	Ar-Ar dating
烟灰色	smoky gray
烟煤	bituminous coal
烟水晶	smoky quartz
胭脂红	carmine
延[展]性	ductility
延长	elongation
延长符号	elongation sign
延展系	ectasian
延展性	ductility
岩鞍	phacolith
岩层	layer
岩层的产状	attitude of stratum
岩层厚度	thickness of stratum
岩床	sill
岩盖	laccolith
岩化[作用]	lithification,diagenesis
岩基	batholith
岩基成矿说	batholith hypothesis of mineralization
岩浆	magma
岩浆成矿作用	magmatic ore-forming process
岩浆-大气水水热系统	magmatic-meteoric hydrothermal system
岩浆房	magma chamber, magma reservoir
岩浆分异作用	magmatic differentiation
岩浆贯入型矿床	magmatic injection-type deposit
岩浆矿床	magmatic mineral deposit

岩浆期后矿床	post-magmatic mineral deposit
岩浆期后热液	postmagmatic hydrothermal solution
岩浆通道	magma conduit
岩浆晚期分异型矿床	late magmatic differentiation-type mineral deposit
岩浆再生[作用]	palingenesis
岩浆作用	magmatism
岩颈	neck
岩脉	vein
岩漠	rocky desert
岩盆	lopolith
岩墙	dyke，dike
岩溶	karst
岩溶洞	karst cave
岩溶湖	karst lake
岩溶水	karst water
岩塞	plug
岩省	rock province
岩石	rock
岩石成因论	petrogenesis
岩石的产状	occurrence of rock
岩石地层单位	lithostratigraphic unit
岩石地球化学测量	goechemical rock survey
岩石化学	petrochemistry
岩石圈	lithosphere
岩石圈板块	lithosphere plate
岩石圈运动	lithosphere movement
岩石异常	rock anomaly
岩体	lithosome
岩席	sheet
岩相	lithofacies
岩相学	petrography

岩屑	lithic fragment, lithic shards, debris, lithoclast, lithic pyroclast
岩屑长石砂岩	lithic feldspar sandstone
岩屑砂屑岩	litharenite
岩屑砂岩	lithic sandstone
岩屑石英砂岩	lithic quartz sandstone
岩屑杂砂岩	lithic wacke
岩性圈闭	lithologic trap
岩盐	halite
岩汁	ichor
岩枝	apophysis
岩钟成矿说	cupola hypothesis of mineralization
岩株	stock
沿岸流	alongshore
研磨材料	abrasive materials
盐	salt
盐度	salinity
盐分	salinity
盐湖	salt lake
盐湖矿床	saline lake deposit
盐湖型锂矿床	salt-lake-type lithium deposit
盐湖型硼矿床	salt-lake-type boron deposit
盐类矿床	salt deposit
盐坪	salt flat
盐丘	salt dome
盐丘圈闭[构造]	salt dome trap[structure]
盐水	brine
盐滩	salt flat, sabkha
盐沼	sabkha
盐渍化	salinization
颜色	color
颜色指数	color index
眼球状构造	augen structure

眼球状混合岩	augen migmatite,ophthalmite
眼球状片麻岩	augen gneiss
演化	evolution
演化线	development line
厌氧过程	anaerobic process
艳蓝色	cobalt blue
堰塞湖	barrier lake
羊背石	sheepback rock
羊膜卵	amnion egg
阳离子	cation
阳起石	actinolite
阳起石化	actinolitization
阳起石片岩	actinolite schist
洋岛玄武岩	oceanic-island-basalt(OIB)
洋底变质[作用]	ocean floor metamorphism
洋底变质岩	ocean-floor metamorphic rock
洋底岩石圈	suboceanic lithosphere
洋红	carmine
洋流	ocean current,sea current
洋隆	oceanic rise
洋中脊	mid-oceanic ridge,mid-ocean ridge
洋中脊变质岩	ocean ridge metamorphic rock
洋中脊岩石	mid-ocean ridge rock
氧	Oxygen(O)
氧黑云母	oxybiotite
氧化	oxidation
氧化带	oxidized zone
氧化还原电位	EH(oxidation-redution potential)
氧化环境	oxidizing environment
氧化物	oxide
氧角闪石	oxyhornblende
氧逸度	oxygen fugacity
样本	specimen

样品	specimen
摇摆石	roggan,rocking stone
耀斑	flare spot
也门	Yemen
野外 P－T 曲线	field P-T curve
野外变质梯度	metamorphic field gradient
野外实习	field practice
叶腊石矿床	pyrophillite deposit
叶蜡石	pyrophyllite
叶蜡石化	pyrophyllitization
叶绿泥石	penninite,pennine
叶绿素体	chlorophyllinite
叶钠长石	cleavelandite
叶片	lamella,blade
叶片状	foliated
叶片状构造	leaf-like structure
叶片状结构	leaf-like texture
叶蛇纹石	antigorite
叶状的	fissile
页岩	shale
液化流	liquefied flows
液化天然气	liquified natural gas(LNG)
液态不混溶作用	liquid immiscibility
液相线	liquidus
一般[单]形	general form
一级相变	first order transition
一水化合物	monohydrate
一致年龄	concordant age
一轴晶	uniaxial crystal
伊拉克	Iraq
伊朗	Iran
伊利诺伊州(美)	Illinois
伊利石	illite

伊利石结晶度	illite crystallinity
衣色蓝	icyl blue
铱	Iridium(Ir)
铱异常	iridium anomaly
仪器中子活化分析	Instrumental Neutron Activation Analysis (INAA)
夷平[作用]	planation
夷平面	planation surface,graded surface
沂蒙矿	yimengite
以色列	Israel
钇	Yttrium(Y)
异常衬度	anomaly contrast
异常干涉色	abnormal interference color
异常规模	anomaly dimension
异常铅	anomalous lead
异常强度	anomaly intensity
异常上地幔	anomalous upper mantle
异常衰减模式	anomaly decay pattern
异常下限	threshold
异地生成煤	allochthonous coal
异化粒	allochem
异化学变质[作用]	allochemical metamorphism
异极	heteropolar
异极矿	hemimorphite
异极象	hemimorphism
异价类质同象	heterovalent isomorphism
异向性	anisotropy
异型	allotype
异源的	allogenetic
易变辉石	pigeonite
易分裂的	fissile
逸度	fugacity
意大利	Italy

意大利蓝	Italian blue
翼[部]	limb
翼间角	interlimb angle
镱	Ytterbium(Yb)
因素	agent
阴丹士林蓝	indanthrene blue
阴极发光	cathodeluminescence
阴极发光显微镜	Cathodoluminescence Microscopy (CLM)
阴离子	anion
阴影状构造	nebulitic structure
铟	Indium(In)
铟矿床	indium deposit
银	Silver(Ag)
银[白]色	silver
银道面	galactic plane
银河[系]	Galaxy, the Milky Way
银河年	galactic year
银金矿	electrum
银矿床	silver deposit
吲哚因蓝	indoine blue
隐蔽圈闭	subtle trap
隐伏矿	buried ore
隐伏矿体	concealed ore body
隐晶集合体	cryptocrystalline aggregate
隐晶质	cryptocrystalline, aphanitic
隐生宙(宇)	Cryptozoic Eon(Eonothem)
印第安纳州(美)	Indiana
印度	India
印度尼西亚	Indonesia
印染蓝	printing blue
印刷蓝	printing blue
印支期	Indosinian
应变场	strain field

应变椭球	strain ellipsoid
应变主轴	principal axes of strain
应变状态	state of strain
应力	stress
应力差	stress difference
应力场	stress field
应力矿物	stress mineral
应力状态	state of stress
应用地球化学	applied geochemistry
应用矿物学	applied mineralogy
英安岩	dacite
英国	United Kingdom
英云闪长岩	tonalite
英云闪长质片麻岩	tonalitic gneiss
英属维尔京群岛	British Virgin Islands
樱草黄	primrose yellow
荧光	fluorescence
荧光显微镜	Fluorescence Microscopy (FM)
萤石	fluorite
萤石矿床	fluorite deposit
萤石型铀矿	fluorite type U-ore
营力	agent
营养级	trophic level
硬度	hardness
硬度测试	Hardness Testing (HDT)
硬腐泥	saprocol
硬绿泥石	chloritoid
硬锰矿	psilomelane
硬模红外光谱	Hard Mode Infrared Spectroscopy (HMIS)
硬砂岩	greywacke
硬石膏	anhydrite
硬水铝石	diaspore
硬玉	jadeite

硬柱石	lawsonite
硬柱石-钠长石-绿泥石相	lawsonite-albite-chlorite facies
优先成核	preferential nucleation
优选方位	preferred orientation
犹他州(美)	Utah
油藏	oil pool
油藏含油高度	oil column height
油层	reservoir
油浸法	immersion method
油苗	oil seepage
油气界面	oil-gas contact
油砂	oil sand
油水界面	oil-water contact
油田	oil field
油页岩	oil shale
油脂光泽	greasy luster,oily luster
油柱高度	oil column height
铀	Uranium(U)
铀-[钍]-铅计时	U-[Th]-Pb dating
铀地球化学旋回	geochemical cycle of uranium
铀-汞型矿化	mineralization of U-Hg type
铀黑	uranium black
铀还原带	uranium reduction zone
铀活化	uranium mobilization
铀矿床	uranium deposit
铀矿地质学	uranium geology
铀-镭平衡	U-Ra equilibrium
铀-磷型矿化	mineralization of U-P type
铀石	coffinite
铀-钛型矿化	mineralization of U-Ti type
铀-铁型矿化	mineralization of U-Fe type
铀系计时	uranium series dating
铀氧化带	uranium oxidized zone

铀氧化-还原过渡带	redox transitional zone of uranium
有机地球化学	organic geochemistry
有机分子起源	origin of organic molecules
有机含硫化合物	sulfur-containing organic compound
有机物残渣	debris
有机质成熟度	maturity of organic matter
有机质热模拟	thermo-simulation of organic matter
有孔虫类	foraminifera
有理指数定律	law of rational indices
有理指数晶面	rational face
有气孔的	vesicular
有色金属矿床	nonferrous metal deposit
有效惰性组分	virtual inert components
有效厚度	net pay thickness
有效渗透率	effective permeability
有序	order
有序参数	order parameter
有序结构	order structure
有序-无序转变	order-disorder transformation
有意义异常	significant anomaly
铕	Europium(Eu)
铕异常	Eu anomaly
黝方石	nosean
黝帘石	zoisite
黝铜矿	tetrahedrite
右形	right-handed form
右旋晶体	right-handed crystal
诱发地震	induced earthquake
余震	aftershock
鱼眼石	apophyllite
宇	Eonothem
宇宙	cosmos，universe
宇宙尘	cosmic dust

宇宙成因核素	cosmogenic nuclide
宇宙大爆炸	big bang
宇宙地质学	space geology, cosmic geology
宇宙丰度	cosmic abundance
宇宙化学	cosmochemistry
宇宙颗粒	cosmic spherule
宇宙矿物学	cosmic mineralogy
宇宙年代学	cosmochronology
宇宙年龄	age of universe
宇宙起源说	cosmogonic theory
宇宙射线	cosmic rays
宇宙线暴露年龄	cosmic ray exposure age
宇宙原子核合成	nucleosynthesis in universe
宇宙源成矿说	cosmogenic hypothesis of minerallization
羽状交错层理	herringbone cross-bedding
雨	rain
雨痕	rain mark, raindrop print
雨水	meteoric waters
玉石矿床	jade and nephrite deposit
玉髓	chalcedony
玉髓化	chalcedonization
玉髓状方解石	brunnerite
育空地区(加)	Yukon Territory
预报	prediction
预防	prevention
预探井	preliminary prospecting well
元古宇	Proterozoic(PR)
元古宙(宇)	Proterozoic Eon(Eonothem)
元素	element
元素比值	element ratio
元素比值图	diagram of element ratio
元素存在形式	mode of occurrence of element, existing form element

元素地球化学	geochemistry of element
元素地球化学分类	geochemical classification of element
元素地球化学行为	geochemical behaviour of element
元素对	element pair
元素分布	distribution of element
元素分布频率	frequency Distribution of element
元素分布图	distribution diagram of element
元素分散	dispersion of element
元素丰度	element's abundance
元素丰度	abundance of element
元素丰度变化图	variation diagram of element abundance
元素富集	enrichment of element
元素合成理论	theory of nucleosynthesis
元素活动性	mobility of element
元素活化	element reactivation
元素亏损	depletion of element
元素年龄	age of element
元素凝聚	condensation of elements
元素浓度	concentration of element
元素起源	origin of element
元素迁移	element transfer，migration of element
元素置换	element substitution
元素组合	element association
原地	situ
原地的	in situ
原地生成煤	autochthonous coal
原地岩[体]	autochthon
原地岩系	autochthon
原恒星	protostar
原钾霞石	kalsilite
原生包体	pramiry inclusion
原生沉积构造	primary sedimentary structure
原生环境	primary environment

原生节理	primary joint
原生矿物	primary mineral
原生铅	primeval lead
原生生物	protist
原生双晶	primary twin
原生异常	primary anomaly
原生晕	primary halo
原始地壳	original crust of earth
原始地幔	primitive mantle(PM)
原始铅	primordial lead
原始水圈	protohydrosphere
原始汤	primeval soup
原始同位素组成异常	primitive isotopic anomaly
原岩	protolith
原油	crude oil
原植物	thallus
原子克拉克值	Clarke of atom
原子力显微镜	Atomic Force Microscopy (AFM)
原子体积	atomic volume
原子吸收光谱	Atomic Absorption Spectroscopy (AAS)
圆度	roundness
圆弧褶皱	arcual fold
圆石	boulder
圆形的	rounded
圆柱状褶皱	cylindrical fold
圆锥颗石	coccolithophora
远成热液矿床	telethermal deposit
远火山活动金属矿床	distal ore deposit
远景区	prospect
远景资源	hypothetical resource
远洋(海)的	pelagic
约旦	Jordan
月光石	moonstone

月海	mare
月核	lunar core
月壳	lunar crust
月陆	lurain
月幔	lunar mantle
月球	moon
月球地质学	lunar geology
月球构造学	lunar tectonics
月球粒陨石	lunar meteorite
月球岩石圈	lunar lithosphere
月壤	lunar soil,lunar regolith
月牙湖	crescent lake
月岩	lunar rock
月震	moonquake
跃移[作用]	saltation [transport]
越南	Viet Nam
云	cloud
云煌岩	minette
云母	mica
云母矿床	mica deposit
云母片岩	mica schist
云母试板	mica plate
云染片麻岩	nebulitic gneiss
云染岩	nebulite
云染状构造	nebulitic structure
云染状混合岩	nebulite
云斜煌岩	kersantite
云英岩	greisen
云英岩化	greisenization
云英岩型锡矿床	greisen-type tin deposit
允许环境极限	acceptable environment limit
允许剂量	acceptable dose
允许浓度	acceptable concentration

陨[石撞]击坑	impact crater
陨[石撞]击作用	meteorite impact
陨击构造	meteorite-impact feature
陨击角砾岩	suevite
陨击坑	crater
陨硫铁	troilite
陨氯铁	lawrencite
陨石	meteorite
陨石坑	meteorite crater
陨石落地年龄	terrestrial age
陨石球粒	chondrule
陨石学	meteoritics
陨石雨	meteorite full
运动学	kinematics
晕彩	iridescence,peristerescence
晕长石	peristerite
晕色	iridescence,peristerescence

Z

杂卤石矿床	polyhalite deposit
杂砂岩	greywacke,graywacke
杂岩	complex
杂质组分	impure components
灾变论	catastrophism
载玻片	slide
载体介质	mounting medium
再沉积作用	redeposition
再活化[作用]	remobilization
再结晶	recrystallization
再区分	redistribute
再生成形的	neomorphic
再生花岗岩	regenerated granite
再生矿床	regenerated deposit

赞比亚	Zambia
藏蓝	purplish blue
藏青	navy blue
糟化	saussuritization
早期成岩作用	early diagenesis
早期岩浆分凝型矿床	early magmatic segregation-type deposit
早期岩浆矿床	early magmatic deposit
藻碧玉	algal jasper
藻鲕	algal ooid
藻类体	alginate
藻煤	boghead coal
藻团	algal lump
藻团粒	algal pellet
藻席	algal mat
藻烛煤	torbanite
造陆运动	epeirogenetic movement
造陆作用	epeirogeny
造山变质[作用]	orogenic metamorphism
造山带	orogenic belt
造山系	Orosirian
造山作用(运动)	orogeny
造型原料矿产	raw material for moulding
造岩矿物	rock-forming mineral
造岩元素	rock-forming element
增田-科里尔图解	Masuda-Coryell diagram
增温层	increasing zone of subsurface temperature
渣块熔岩	aa lava
扎布耶石	zabuyelite
扎伊尔	Zaire
乍得	Chad
展性	malleability
张衡矿	zhanghengite
张节理	tension joint

张量	tensor
章氏硼镁石	hungchaoite
涨滩	alluvion
涨湾	embayment
障蔽坝	barrier bar
障蔽滩	barrier beach
障壁岛	barrier island, barrier
找矿标志	clue for prespecting
找矿矿物学	prospecting mineralogy
沼泽	marsh, bog, swamp, mire
照片	picture
遮光器	dimmer
折光率	refractive index
折射	refraction
折射计	refractometer
折射率	refractive index
折射仪	Abbe refractometer
哲兹卡兹甘铜矿床	Dzhezkazgan copper deposit
锗	Germanium(Ge)
锗矿床	germanium deposit
赭[石]色	sienna, chocolate
赭土	umber
褶皱	fold
褶皱[作用]	folding
褶皱波长	wavelength
褶皱分析	fold analysis
褶皱干扰型式	fold interference pattern
褶皱构造地貌	fold geomorphology
褶皱山	folded mountain
褶皱世代	fold generation
褶皱要素	elements of fold
褶皱轴	axis of fold
针沸石	mazzite

针钠钙石	pectolite
针铁矿	goethite
针状	acicular(needle)
珍珠光泽	pearly luster
珍珠岩	perlite
珍珠岩矿床	perlite deposit
珍珠云母	margarite
真等变线	true isograd
真鲕	true ooid
真核生物	eucaryote
枕状构造	pillow structure
枕状熔岩	pillow lava
震旦纪(系)	Sinian Period(System)
震源	seismic focus, seismic source, hypocenter
震源深度	focal depth, seismic source depth
震中	epicenter
震中烈度	epicentral intensity
蒸发	evaporate
蒸发沉积矿床	evaporite deposit
蒸发岩	evaporite
蒸发岩矿床	evaporite depisit
蒸发盐	evaporite
蒸汽压	vapor pressure
整合	conformity
整合侵入体	concordant intrusion
整数定律	law of rational indices
正变基性岩	orthometabasite
正变质岩	orthometamorphic rock, orthometamorphite
正长片麻岩	orthoclase gneiss
正长石	orthoclase
正长石化	orthoclasization
正长岩	syenite

正常层序	normal succession, right way up
正常火山碎屑岩	ordinary pyroclastic rock
正断层	normal fault
正方形格子	square grid
正格子(与倒易格子对应)	direct lattice
正光性	positive character
正硅酸盐	orthosilicate
正交晶系	orthorhombic system
正交偏光	cross-polarized light(cpl)
正交剖面图	profile
正交双晶	normal twin
正角闪岩	orthoamphibolite
正片麻岩	orthogneiss
正片岩	orthoschist
正突起	positive relief
正烷烃分布	distribution of n-alkanes
正烷烃奇偶优势	odd-even predominance of n-alkanes
正向期	normal epoch
正形	positive form
正延性	length slow, positive elongation
正岩浆矿床	orthomagmatic mineral deposit
正异常	postive anomaly
正应力	normal stress
支撑类型	supporting style
支地槽	embayment
直立褶皱	upright fold
直闪石	anthophylite
直线滑移	translation gliding
直消光	straight extinction, parallel extinction
植物根痕	root marks
指标元素	indicator element
指示矿物	index mineral
指示矿物带	index mineral zone

指示植物	indicator plant
志留纪(系)	Silurian Period(System),Silurian
质量监控	data quality monitoring
质谱仪	Mass-Spectroscopy (MS)
质子发光	proton luminescence
质子诱导 X 射线发射光谱	Proton-Induced X-ray Emission Spectroscopy (PIXES)
质子诱导 γ 射线发射光谱	Proton-Induced Gamma-ray Emission Spectroscopy (PIGME)
栉壳状结构	ctenoid texture
智利	Chile
智利硝石矿床	soda niter deposit, nitratine deposit
滞留颗粒	lag grains
滞水	perch ground water
蛭石	vermiculite
蛭石矿床	vermiculite deposit
置换	substitution
中奥陶统	Middle Ordovician
中长石	andesine
中带	mesozone
中等解理	fair cleavage
中非共和国	Central African Republic
中非铜矿床带	Central-African copper belt
中沸石	mesolite
中国	China
中寒武统	Middle Cambrian
中和反应	neutralization
中和作用	neutralization
中级变质岩	medium-grade metamorphic rock
中键结构	mesodesmic structure
中砾	pebble
中砾岩	pebble conglomerate
中粒长石杂砂岩	medium arkose

中粒的	medium grained
中粒结构	medium grained texture
中泥盆统	Middle Devonian
中期预报	medium-range forecast
中三叠统	Middle Triassic
中色的	mesocratic
中色体	mesosome
中色岩	mesocratic rock
中砂岩	medium sandstone
中生代(界)	Mesozoic Era(Erathem), Mesozoic(MZ)
中太古界	Mesoarchean
中条纹长石	mesoperthite
中铁陨石	mesosiderite, pyroxene-plagioclase, stony-iron
中突起	median relief
中温热液矿床	mesothermal deposit
中温水热矿床	mesothermal deposit
中心喷发	central eruption
中心式喷发	central eruption
中新世(统)	Miocene Epoch(Series), Miocene
中性	Intermediate
中性蓝	neutral blue
中性麻粒岩	intermediate granulite
中性岩	intermediate rock
中压相系	medium-pressure facies series
中元古代(界)	Mesoproterozoic Era(Erathem), Mesoproterozoic
中源地震	intermediate-focus earthquake
中侏罗统	Middle Jurassic
中柱石	Mizzonite
中子粉末衍射	Neutron Powder Diffraction (NPD)
中子活化分析	Neutron Activation Analysis (NAA)
中子星	neutron star

中子衍射	Neutron Diffraction (ND)
终端反应	terminal reaction
终碛堤	end moraine bar
钟乳状集合体	stalactitic aggregate
种(矿物分类基本单位)	species
种群	population
重荷模	load casts
重建式相变	reconstructive transformation
重建型转变	reconstructive transformation
重金属污染	heavy metal pollution
重晶石	barite
重晶石矿床	barite deposit
重矿物	heavy mineral
重力	gravity
重力[地质]作用	gravitational process
重力均衡	gravity isostasy
重力式胶结	gravity cementation
重力水	gravity water
重力泄油储油层	gravity drive reservoir
重力异常	gravity anomaly
重膜状矿脉	sheeted-zone veins
重砂矿床	heavy mineral deposit
重砂矿物	placer mineral
重夕线石	xenolite
重稀土元素	heavy rare earth element(HREE)
重质油	heavy oil
周期性键链理论	periodic bond chain (PBC) theory
轴长	axial length
轴迹	axial trace
轴角	axial angle
轴率	axial ratios
轴面	axial surface,axial plane, pinacoid
轴面劈理	axial plane cleavage

轴双面	axial dihedron
轴向分带	axial zoning
轴截距	axial intercept
肘状双晶	elbow twin,knee shaped twin
宙	Eon
皱纹构造	plicate structure,plicated structure
皱纹线理	crenulation lineation
朱红	vermilion,vermeil,minium
侏罗纪(系)	Jurassic Period(System),Jurassic
竹绿色	green blue
烛煤	cannel coal
主量元素	bulk element,major element
主色调	hue
主星系	main sequence
主序带	main sequence
主岩	host rock
主应力	principal stress
主元素	major element
主折射率	principle refractive index
主震	main shock
注水[油]井	water injection well
柱	prism
柱面双晶	prism twin
柱状	columnar,prismatic
柱状集合体	columnar aggregate
柱状节理	columnar jointing
铸模	casts,molds
铸模孔	mold pore
铸造材料	founding materials
砖红土	laterite,latosol
转变双晶	transformation twin
转换断层	transform fault
转换断层型边界	transform boundary

转折端	hinge zone
撞击角砾岩	suevite
撞击坑	impact crater
锥	pyramid
锥辉石	acmite
锥形的	tapered
锥状褶皱	conical fold
准等变线	tentative isograd
准晶态	quasicrystal
准晶体	quasicrystal
准块状构造	para-massive structure
准矿物	mineraloid
准平原	peneplain
准平原化[作用]	peneplanation
准同生作用	penecontemporaneous
准稳[定状]态	metastable state
准稳的	metastable
浊沸石	laumontite
浊积岩	turbidite
浊流	turbidity current
浊流层	turbidite
资源	resources
资源的变动性	motility of resources
资源的地域性	regionalization of resources
资源的多用性	multi-utility of resources
资源的社会性	sociality of resources
资源的稀缺性	scarcity of resources
资源的整体性	integrity resources
资源短缺	resource shortage, resource scarcity
资源概念的动态性	dynamics of resource concept
资源基础	resource base
资源评价	resource assessment
子群	subgroup

紫硅镁铝石	yoderite
紫褐色	purple brown
紫黑色	purple black
紫红	mauve, violet red, violet carmine, purplish red
紫灰色	purplish grey
紫蓝色	purplish blue, indigo blue, thumb blue
紫罗兰色	pansy, violet
紫色	purple
紫水晶	amethyst
紫苏花岗岩	charnockite, hypersthene granite
紫苏辉石	hypersthene
紫外高真空扫描隧道显微镜	Ultraviolet High-Vacuum Scanning Tunneling Microscopy (UHV STM)
紫外光电子光谱	Ultraviolet Photoelectron Spectroscopy (UPS)
紫外激光磨蚀质谱仪	Ultraviolet Laser Ablation Mass Spectroscopy (UV-LA-MS)
紫外可见光谱	UV-Visible Spectroscopy (UV-VS)
紫棕色	purple brown
自变质岩	autometasomatic rock
自催化作用	autocatalysis
自发裂变	spontaneous fission
自扩散作用	self-diffusion
自然铋	bismuth
自然铂	platinum
自然残留磁性测量	Natural Remnant Magnetism Measurements (NRMM)
自然地带性	phisical zonality
自然地理环境	physic-geographical environment
自然地理系统	physic-geographical system
自然铬	native chromium
自然环境	natural environment

自然金	gold
自然硫	sulphur
自然硫矿床	native sulfur deposit
自然释放	natural release
自然铜	copper
自然灾害	natural disaster, hazard
自然资源	natural resource
自然资源的可持续利用	sustainable use of natural resources
自色	idiochromatic color
自生沉积物	authigenic sediment
自生沉积岩	authigenic sedimentary rock
自生成因	authigenesis
自生矿物	authigenic mineral
自生碎屑	autoclast
自生作用	authigenesis
自碎作用	autoclasis
自我平衡	homeostasis
自限性	selfconfinement
自形变晶	idioblastic
自形粒状结构	idiomorphic granular texture
自型	automorphic
自养生物	autotrophic organism
自转	autogiration
棕褐色	tawny, dark brown, tan
棕灰色	brownish gray
棕色	umber, brown
总初级生产量	gross primary production
总滑距	net slip
总鳍鱼类	Crossopterygii
总星系	metagalaxy
纵波(P 波)	longitudinal wave(P wave)
纵弯褶皱	buckle fold
纵向分带	longitudinal zoning

纵向沙丘	longitudinal dune
走时	travel time
走向	strike
走向断层	strike fault
走向滑动断层	strike-slip-fault
足印(迹)	footprint, tracks
族(矿物分类单位)	group
组	formation
组分差异活动性	differential mobility of components
组构	fabric
祖母绿	emerald
钻井	borehole
钻孔生物	burrowing organisms
钻石	diamond
钻石高压盒	Diamond Anvil Cell (DAC)
最大主应力	maximum principal stress
最大资源潜力	maximum resource potential
最佳估计值	best estimate
最小内能性	minimum internal energy
最小主应力	minimum principal stress
最终可采资源	ultimately recoverable resource
左形	left-handed form
左旋晶体	left-handed crystal
左右形关系	enantiomorphic relationship
佐治亚州(美)	Georgia
作图	construction
作用力	agent
坐标方向	coordinate direction

2. 英汉地球物质科学术语

English-Chinese Earth Material Science Vocabularies

A

a light bright yellow	藤黄(浅亮黄色)
A'KF diagram	A'KF 图
aa lava	渣块熔岩
AAGS	非洲地质调查所协会
AAS	原子吸收光谱
Ab initio	一种计算机模拟方法
Abbe refractometer	阿贝折射仪
abnormal interference color	异常干涉色
abrasion	磨蚀[作用]
abrasive materials	研磨材料
absinthe-green	淡绿色
absolute abundance	绝对丰度
absorbent charcoal method	活性碳法
absorption	吸收性
absorption index	吸收指数
Absorption Spectroscopy (AS)	吸收光谱
absorptive capacity	吸收能力
absorptivity	吸收性
abstraction	退流
abundance	丰度
abundance of element	元素丰度
abyssal	深海的
AC	器官系数
Accelerated Mass-Spectroscopy (AMS)	加速质谱仪
acceptable concentration	允许浓度

acceptable dose	允许剂量
acceptable environment limit	允许环境极限
accessory mineral	副矿物
acclimation disease	水土病
accommondation	可容纳空间
accumulation	堆积[作用]
accumulative action	累积作用
accumulative coefficient	累积系数
accumulative factor	累积因子
ACF diagram	ACF 图
achondrite	无球粒陨石
acicular	针状
acid clay	酸性白土
acid environment	酸性环境
acid precipitation	酸雨
acid rain	酸雨
acidic	酸性的
acidic rock	酸性岩
acline twins	底面双晶,水平双晶
acmite	锥辉石
ACMRR	海洋资源研究咨询委员会
acropetal coefficient(AC)	器官系数
Actinium(Ac)	锕
actinolite	阳起石
actinolite schist	阳起石片岩
actinolitization	阳起石化
activation-type orogeny	活动型造山作用
active acidity	活性酸度
active tectonic belt	活动构造带
active tectonics	活动构造
active volcano	活火山
actualism	现实主义[原则和方法]
adakite	埃达克岩

adamantine luster	金刚光泽
adamite	羟砷锌石
additive components	加成组分
additive halo	累加晕
adiabatic compression	绝热压缩
adsorption	吸附作用
adsorption uranium	吸附铀
adular	冰长石
adularescence	冰长石晕彩
adularia	冰长石
adularization	冰长石化
Advisory Committee on Marine Resources (ACMRR)	海洋资源研究咨询委员会
aegirine	霓石
aegirine-augite	霓辉石
aegrinization	霓石化
aeolian	风成的
aeolian basin	风蚀盆地
aeolian sediment	风积物
aeolian valley	风蚀谷
aerobic process	亲氧过程
Afghanistan	阿富汗
AFM	原子力显微镜
AFM diagram	AFM 图
aftershock	余震
Age	年代
age dating	计时
age determination	年龄测定
age of element	元素年龄
age of gymnosperm	裸子植物时代
age of universe	宇宙年龄
age spectrum	年龄谱
ageing	陈化,老化

agent	营力,作用力,因素,媒介物,介质
agglomerate	集块岩
aggradation	填积作用,加积作用,叠积作用
aggradational	层序加积的
aggregate	集合体,集料
aggregated element	聚集元素
AGID	国际发展地学家协会
agitate	搅动
agitated leaching	搅动浸出
agmatite	角砾状混合岩
Air Comparison Pycnometer	空气对比比重仪(比重瓶)
air mass	气团
air temperature	气温
air temperature lapse rate	气温递减率
airborne debris	散落物
airborne geochemical exploration	航空化探
Alabama	亚拉巴马州(美)
Alaska	阿拉斯加州(美)
alaskite	白岗岩
Albania	阿尔巴尼亚
Alberta	艾伯塔省(加)
albite	钠长石
albite law	钠长石律双晶
albite-epidote hornfels	钠长-绿帘角岩
albite-epidote hornfels facies	钠长-绿帘角岩相
albite-gneiss	钠长片麻岩
albitization	钠长石化
algal jasper	藻碧玉
algal lump	藻团
algal mat	藻席
algal ooid	藻鲕
algal pellet	藻团粒
Algeria	阿尔及利亚

alginate	藻类体
Algoma-type iron formation	阿尔戈马型含铁建造
aliphatic fraction	烷烃馏分
alizarin red	茜素红
alkali basalt	碱性玄武岩
alkali feldspar	碱性长石
alkali rock	碱性岩
alkalic-metasomatism type uranium deposit	碱交代型铀矿
alkali-feldspar gneiss	碱长片麻岩
alkaline amphibolization	碱性角闪石化
alkaline earths	碱土
alkaline environment	碱性环境
alkaline metasomatism	碱交代[作用]
alkaline pyroxenization	碱性辉石化
alkaline rock type uranium deposit	碱性岩型铀矿
alkaline series	碱性系列
alkalis	碱
alkyl benzenes	烷基苯系列化合物
alkyl biphenyls	烷基联苯化合物
alkyl naphthalenes	烷基萘系列化合物
alkyl phenanthrenes	烷基菲系列化合物
allanite	褐帘石
allochem	异化粒
allochemical metamorphism	异化学变质[作用]
allochromatic color	他色
allochthon	外来[岩]体
allochthonous coal	异地生成煤
allogenetic	异源的,外源的
allogenic mineral	他生矿物
allogenic rock	他生沉积岩
allogenic sedimentary rock	他生沉积岩
allomerism	类质同象
allotrioblastic	它形变晶[的]

allotriomorphic granular	他形粒状
allotriomorphic granular texture	他形粒状结构
allotype	他型,异型
alloy	金属互化物
alluvial fan	冲积扇
alluvial placer	冲积砂矿
alluvial plain	冲积平原
alluvion	冲积层,涨滩,海浪冲刷
almandite	铁铝石榴子石
alongshore	沿岸流
Alpine	阿尔卑斯期
alteration	改造,蚀变,交替,互层
altered rock	蚀变岩石
altered wallrock	蚀变围岩
altitude effect	高度效应
altitudinal zonality	垂直分带性
aluminium deposit	铝矿床
Aluminium(Al)	铝
aluminosilicate	铝硅酸盐
aluminous rocks	铝质岩
alunite	明矾石
alunite deposit	明矾石矿床
alunitization	明矾石化
amazonite	天河石
amazonitization	天河石化
amber	琥珀[黄]色
Americium(Am)	镅
amethyst	紫水晶
amino acid	氨基酸
amino acids	氨基酸类化合物
amino acids age determination	氨基酸定年
amnion egg	羊膜卵
amorphous	非晶质的(无定形的)

amorphous substance	非晶质,玻璃质
amount of precipitation	降水量
Amphibia	两栖类
amphibole	角闪石
amphibolite	角闪岩
amphibolite facies	角闪岩相
AMS	加速质谱仪
amygdaloid	杏仁体
amygdaloidal structure	杏仁构造
anabolism	合成作用,同化作用
anaerobic process	厌氧过程
analbite	高温钠长石
analcidite	方沸石
analcime	方沸石
analcite	方沸石
anatexis	深熔[作用]
anchi-metamorphic rock	近变质岩
andalusite	红柱石
andalusite-sillimanite type(or series)	红柱石-夕线石型(系列)
andesine	中长石
andesite	安山岩
andesitic	安山质
Andorra	安道尔
andradite	钙铁石榴子石
angle of polarization	偏振角
Angle- Resolved X-ray Photoelectron Spectroscopy (ARXPS)	角度分辨X射线光电子谱仪
Angola	安哥拉
Anguilla	安圭拉岛(英)
angular	棱角状的
angular unconformity	角度不整合
anhydrite	硬石膏,无水石膏
anion	阴离子

anisodesmic structure	非均键结构
anisotropic body	非均质体
anisotropy	各向异性,异向性
ankangite	安康矿
ankerite	铁白云石
annite	羟铁云母
anomalous lead	异常铅
anomalous upper mantle	异常上地幔
anomaly contrast	异常衬度
anomaly decay pattern	异常衰减模式
anomaly dimension	异常规模
anomaly intensity	异常强度
anorthite	钙长石
anorthoclase	歪长石
anorthosite	斜长岩
Anshan-type of iron deposit	鞍山式铁矿床
antecedent river	先成河
anthophylite	直闪石
anthracite	无烟煤
anthropogenic discharge	人为排放
anticline	背斜
anticline theory	背斜构造理论
anticlinorium	复背斜
anticyclone	反气旋
antiferromagnetism	反铁磁性
antiform	背形
antigorite	叶蛇纹石
Antigua and Barbuda	安提瓜和巴布达
antimony deposit	锑矿床
antimonite	辉锑矿
Antimony(Sb)	锑
antiperthite	反条纹长石
antiphase domains	反相畴

antique violet	古紫色
antistress mineral	反应力矿物
apatite	磷灰石
apatite deposit	磷灰石矿床
apatitolite	磷灰石岩
apex	顶
aphanitic	隐晶质
aplite	细晶岩
aplitic texture	细晶结构
apophyllite	鱼眼石
apophysis	岩枝
apparent age	表观年龄
apparent dip	视倾斜,视倾角
apparent geothermal gradient	视地热梯度
apple green	苹果绿
applied geochemistry	应用地球化学
applied mineralogy	应用矿物学
appraisal well	评价井,详探井
aqua	浅绿色
aquamarine blue	蓝绿色
aquifer	含水层
aragonite	文石,霰石
Ar-Ar dating	氩-氩计时
Archaean	太古宇(宙)
Archaean Eon(Eonothem)	太古宙(宇)
Archean nucleus	太古代陆核
arc-trench gap	弧-沟间隙
arcual fold	圆弧褶皱
areal geochemical anomaly	区域地球化学异常
areal productivity	面金属量
arenite	净砂岩,静砂岩
arfvedsonite	钠铁闪石(又称:亚铁钠闪石)
Argentina	阿根廷

argentite	辉银矿
argillaceous	泥质,粘土质
argillified type U-ore	粘土化蚀变型铀矿
argillite	泥岩,泥板岩,泥质岩,粘土岩
argillization	泥化
Argon(Ar)	氩
aridity index	干燥指数
Arizona	亚利桑那州(美)
Arkansas	阿肯色州(美)
arkose	长石砂岩
Armenia	亚美尼亚
arsenate	砷酸盐
arsenic deposit	砷矿床
Arsenic(As)	砷
arsenopyrite	毒砂
arterite	脉混合岩
artesian water	承压水
artificial mineral	人工矿物,合成矿物
Aruba	阿鲁巴岛
AS	吸收光谱
asbestos deposit	石棉矿床
ascending solution	上升溶液
ascharite deposit	硼镁石矿床
ash	火山灰
ash flow	火山碎屑流
ash texture	凝灰结构
ashing	灰化
asphalt	沥青
assimilation	同化[作用]
assimilation and fractional crystallization (AFC)	同化混染与分离结晶[作用]
associate	伴生
associate mineral	伴生矿物

associated gas	伴生气
associated ore	伴生矿
Association of African Geological Surveys (AAGS)	非洲地质调查所协会
Association of Geoscientists for International Development (AGID)	国际发展地学家协会
association of ore deposits	矿床组合
Astatine(At)	砹
asterism	星彩
asteroid	小行星
asthenosphere	软流圈
astrobleme	古陨击坑
astrochemistry	天体化学
astrogeology	天体地质学
astronomical unit(AU)	天文单位
asymmetrical fold	不对称褶皱
asymmetrical ripple	不对称波痕
asymmetry	非对称
Ata-Su manganese deposit	阿塔苏锰矿床
ataxite	无结构铁陨石
atmophile element	亲气元素,惰性气体
atmosphere	大气圈
atmospheric argon	大气氩
atmospheric pressure	气压
Atomic Absorption Spectroscopy (AAS)	原子吸收光谱
Atomic Force Microscopy (AFM)	原子力显微镜
atomic volume	原子体积
atrous	深黑
attachment kinetics	附着能,吸附能
attapulgite clay deposit	凹凸棒石粘土矿床
attapulgite deposit	凹凸棒石矿床
attitude	产状
attitude of stratum	岩层的产状

AU	天文单位
auduoite	安多矿
augen gneiss	眼球状片麻岩
augen migmatite	眼球状混合岩
augen structure	眼球状构造
augite	普通辉石
auri-uraniferous conglomerate	金-铀砾岩
Australia	澳大利亚
Austria	奥地利
authigenesis	自生作用,自生成因
authigenic mineral	自生矿物
authigenic sediment	自生沉积物
authigenic sedimentary rock	自生沉积岩
autocatalysis	自催化作用
autochthon	原地岩[体],原地岩系
autochthonous coal	原地生成煤
autoclasis	自碎作用
autoclast	自生碎屑
autoclave	高压釜
autogiration	自转
autometasomatic rock	自变质岩
automorphic	自型
autotrophic organism	自养生物
aventurescence	砂金石闪光
aventurine	砂金石(富包体石英)
axial angle	轴角
axial dihedron	轴双面
axial intercept	轴截距
axial length	轴长
axial plane	轴面
axial plane cleavage	轴面劈理
axial ratios	轴率
axial surface	轴面

axial trace	轴迹
axial zoning	轴向分带
axinite	斧石
axinitization	斧石化
axis of fold	褶皱轴
Azerbaijan	阿塞拜疆
azimuth	方位[角]
azonality	非地带性
azurite	蓝铜矿
azure	天蓝
azury	青色,蓝绿色,浅蓝

B

BABI(basaltic achondrite best initial)	BABI 值(玄武质无球粒陨石最佳初始值,代表地球形成时的$^{87}Sr/^{86}Sr$值)
Babinet compensator	巴比涅补偿器
baby blue	淡蓝
baby pink	浅粉红色
BAC	生物吸收系数
back-arc basin	弧后盆地
background extinction	绝灭背景质
background value	背景值
Backscattered Electron Imaging (BEI,BSEI)	背散射电子像
Backscattered Electron Microscopy (BSEM)	背散射电子显微镜
Backscattered Scanning Electron Imagery (BSE)	背散射扫描电子图像
bacteria action	细菌作用
baddeleyite	斜锆石
Bϕggild intergrowth	博吉尔德连生(斜长石出溶形成)
bafertisite	钡铁钛石

Bahamas	巴哈马
Bahrain	巴林
baiyuneboite	白云鄂博矿
baked zone	烘烤带
balanced section	平衡剖面
balipholite	纤钡锂石
ball clay	球土
banalsite	钠钡长石
band gap	带隙
Band theory	能带理论
banded cherty iron formation	条带状硅铁建造
banded differentiate	条带状分异岩
banded hematite quartzite	条带状赤铁石英岩
banded marble	条带状大理岩
banded quartzite	条带状石英岩
banded sediment	带状沉积
banded structure	条带状构造
banding	条带状构造
Bangladesh	孟加拉国
bank reef	岸礁
baotite	包头矿
Barbados	巴巴多斯
barchan	新月形沙丘
barite	重晶石
barite deposit	重晶石矿床
Barium(Ba)	钡
barren bed	哑层
barrier	障壁岛,坝
barrier bar	障蔽坝，滨外沙坝
barrier beach	滨外滩,障蔽滩
barrier island	障壁岛
barrier lake	堰塞湖
barrier reef	堡礁

barroisite	冻蓝闪石
Barrovian zones	巴罗带
barytolamprophyllite	钡闪叶石
basal cementation	基底式胶结
basal pinacoid	底[轴]面
basal section	横切面
basalt	玄武岩
basaltic	玄武质
basaltic andesite	玄武安山岩
basanite	碧玄岩
base level of erosion	侵蚀基准面
base-centered lattice	底心格子(C-格子)
basement	基底
basement maturity	基底成熟度
basic	基性
basic granulite	基性麻粒岩
basic hornfels	基性角岩
basic rock	基性岩
basin	盆地
basin-range structure	盆-岭构造
basite	基性岩
bassanite	烧石膏
bastnaesite	氟碳铈矿
batch melting	批式熔融,间歇熔融
batholith	岩基
batholith hypothesis of mineralization	岩基成矿说
bathyal	半深海的
bathyal sediment	半深海沉积物
bauxite	铝土矿
bauxite deposit	铝土矿矿床
Baveno twin	巴韦诺双晶(曾用名:巴温诺双晶)
bay	海湾

beach	海滨,湖滨,河滩
beach drift	进浪流
beach placer	海滨砂矿
beachrock	滩岩,海滩岩
bearing	方位角
Becke line	贝克线
bed	层,地层
bed separation	层状剥落
bedded vein	似层状矿脉
bedding	成层,层理
bedding plane	层面
bedrock	基岩
bedrock coast	基岩海岸
BEI	背散射电子像
beige	米色,浅褐色
Beking man	北京[直立]人
Belarus	白俄罗斯
Belgium	比利时
Belize	伯利兹
bench placer	阶地砂矿
bending fold	板褶皱,横弯褶皱
Benin	贝宁
Benioff zone	地震带,贝尼奥夫带
benthonic	底栖的,海底生物
bentonite	膨润土
bentonite deposit	膨润土矿床
Berek compensator	贝累克补偿器(曾用名:贝瑞克补色器)
beresite	黄铁绢英岩
beresitization	黄铁绢英岩化
Berkelium(Bk)	锫
Berlin blue	柏林蓝,普蓝
Bermudas	百幕大群岛(英)

Bernal chart	贝尔纳图
bertranditization	硅铍石化
beryl	绿柱石
beryl pegmatite deposit	绿柱石伟晶岩矿床
berylitization	绿柱石化
beryllium deposit	铍矿床
Beryllium(Be)	铍
beryl-quartz vein	绿柱石石英脉
berzeliite	黄砷榴石
best estimate	最佳估计值
Bhutan	不丹
bias	偏倚
biaxial crystal	二轴晶
bifurcation of coal seam	煤层分叉
big bang	宇宙大爆炸,大爆炸
big-bang cosmology	大爆炸宇宙学
bimetasomatism	双交代[作用]
binary eutectic	二元共晶
bioaccumulation	生物累积
biochemical sedimentary rock	生物化学沉积岩
biochemical weathering	生物化学风化作用
bioclast	生物碎屑
biodegradation	生物降解作用
bio-diversity	生物多样性
bioelement	生命元素
biofacies	生物相，化石相
biogenetic gas	生物成因气
biogenic process	生物成因作用
biogenic sediment	生物沉积
biogeochemical cycle	生物地球化学循环
biogeochemical survey	生物地球化学测量
biogeochemistry	生物地球化学
bioherm	生物礁

biological absorption coefficient(BAC)	生物吸收系数
biological adaptation	生物的适应性
biological concentration	生物富集
biological indicator	生物的指示现象
biological marker compound	生物标志化合物
biological purification	生物净化
biological weathering	生物风化
biomarker	生物标志化合物
bio-mass	生物量
biome	生物群落
biomicrite	生物泥晶灰岩,生物微晶灰岩
biomineralization	生物矿化作用
biophile element	亲生物元素
bioslime	生物软泥
biosparite	生物亮晶灰岩
biosphere	生物圈
biostratification	生物地层层理
biostratigraphic unit	生物地层单位
biostratigraphic zone	生物地层带
biostratigraphy	生物地层学
biotic community	生物群落
biotic explosion	生物爆发
biotite	黑云母
biotitization	黑云母化
bioturbation	生物扰动作用,生物搅动作用
bioturbation structure	生物扰动构造
biprism	复柱
bipyramid	双锥
birdseye structure	鸟眼构造
bireflection	双反射
birefringence	双折射率,重折率
bisectrix	等分线
bismuth	自然铋

bismuth deposit	铋矿床
Bismuth(Bi)	铋
bismuthinite	辉铋矿
bisque	桔黄色
bitumen	沥青
bituminite	沥青质体
bituminous coal	烟煤
bituminous shale	沥青质页岩
black	黑色
black hole	黑洞
black quartz	墨晶
black smoker	海底黑烟柱
blade	叶片
blast	变晶
blastic texture	变晶结构
blastobedding	变余层理
blastomylonite	变余糜棱岩
blastopyroclastic texture	变余火成碎屑结构
blastoporphyritic texture	变余斑状结构
blastopsammitic texture	变余砂状结构
blastopsephitic texture	变余砾状结构
blind ore	盲矿
blind ore body	盲矿体
blind valley	盲谷
block diagram	立体图
blossom	煤华
blowout	风蚀洼地
blue	蓝色
blue black	蓝黑色,深蓝色
blue green	青绿的
blue schist	蓝片岩
blueish	带青色的
blueschist	蓝片岩

blueschist facies	蓝片岩相
bluish green	蓝绿色,深绿
body wave	体波
body-centered lattice	体心格子(I-格子)
bog	沼泽
boghead coal	藻煤
Bolivia	玻利维亚
bomb	火山弹
bonanza	大矿囊
bond energy	键能
bond length	键长
bond strength	键强
boracite	方硼石
borax	硼砂
borehole	钻井
bornite	斑铜矿
Born-Mayer theory	玻恩-梅尔理论
boron deposit	硼矿床
Boron(B)	硼
boron-bearing skarn deposit	夕卡岩硼矿床
Bosnia and Hercegovina	波斯尼亚和塞尔维亚
botryoidal aggregate	葡萄状集合体
botryoidal structure	葡萄状构造
Botswana	博茨瓦纳
Bottle Pycnometer	比重瓶
Bouguer gravity anomaly	布格重力异常,布伽重力异常
boulder	漂砾,巨砾,圆石
boulder-clay	泥砾
Bouma sequence	鲍玛序列
bound water	结合水
boundary clay	界限粘土
boundstone	生物粘结灰岩
box fold	箱状褶皱

boxwork	箱状构造物,蜂窝状网格,网格构造
brachiopoda	腕足动物门
brachyfold	短轴褶皱
Bragg equation	布拉格方程
Bragg's law	布拉格定律
braided stream	辫状河流,网状河流,交织河流
brassy yellow	铜黄色
Bravais indices	布拉维指数
Bravais lattice	布拉维晶格
Brazil	巴西
Brazil twin	巴西双晶
breaker	拍岸浪,破浪
breccia	角砾岩
breccia structure	角砾状构造
brecciated structure	角砾状构造
breunnerite	铁菱镁矿
Brewster's law	布儒斯特定律(曾用名:布鲁斯特定律)
bright	明亮的
bright coal	光亮煤
bright red	绯红,深红,鲜红,大红
brine	盐水,卤水
brine deposit	卤水矿床
British Columbia	不列颠哥伦比亚省(加)
British Virgin Islands	英属维尔京群岛
brittle behavior	脆性行为
brittle deformation	脆性变形
brittle fracture	脆性断裂
brittleness	脆性
Bromine(Br)	溴
bronze	青铜色,古铜色,赤褐色,红褐色
bronze blue	青铜蓝,普鲁士蓝

bronze smoke	青铜烟灰色
bronze yellow	青铜黄色的
bronzed	青铜色,古铜色,赤褐色
bronzite	古铜辉石
bronzite-olivine stony-iron	古铜橄榄石铁陨石
brown	褐色,棕色
brown coal	褐煤
brownish	带褐色的
brownish black	褐黑色
brownish gray	棕灰色
brownish green	褐绿色
brucite	水镁石,氢氧镁石
Brunei	文莱
Brunhes polarity chron	布容极性时
brunnerite	玉髓状方解石
bryozoan	苔藓虫,苔藓虫类
BSE	背散射扫描电子图像
BSEI	背散射电子像
BSEM	背散射电子显微镜
bubble	气泡
buchite	假熔岩
buckle fold	弯褶皱,纵弯褶皱
buffer action	缓冲作用
buffer reaction	缓冲反应
buffering	缓冲
buffering agent	缓冲剂
building materials	建筑材料
Bulgaria	保加利亚
bulk element	主量元素
bulk rock	全岩
burial metamorphic rock	埋藏变质岩
burial metamorphism	埋藏变质[作用]
buried ore	隐伏矿

buried ore body	埋藏矿体
buried terrace	埋藏阶地
buried-hill trap[structure]	潜山圈闭[构造]
Burkina Faso	布基纳法索
Burma	缅甸
burrow mottled sediment	穴斑沉积物
burrowing organisms	钻孔生物
burrows	潜穴,生物钻孔 ,虫孔
Burundi	布隆迪
Bushveld chromium deposit	布什维尔德铬矿床
bytownite	培长石

C

cadmium deposit	镉矿床
Cadmium(Cd)	镉
Caesium(Cs)	铯
Cainozoic Era(Erathem)	新生代(界)
Calacas-type iron deposit	卡腊贾斯型铁矿床
calc-gneiss	钙质片麻岩
calc rock	钙质岩
calc-schist	钙质片岩
calc-alkali index	钙碱指数
calc-alkaline series	钙碱性系列
calcareous	钙质
calcareous feldspathic arenite	钙质长石砂岩
calcareous gneiss	钙质片麻岩
calc-gneiss	钙质片麻岩
calcite	方解石
calcitization	方解石化
Calcium(Ca)	钙
calcium-aluminum-rich inclusion	富钙铝包体
calcium-poor achondrite	贫钙无球粒陨石
calcium-rich achondrite	富钙无球粒陨石

calc-phyllite	钙质千枚岩
calc-schist	钙质片岩
calc-silicate fels	钙硅酸盐粒岩
calc-silicate hornfels	钙硅酸盐角岩
calc-silicate rock	钙硅酸盐岩
calc-slate	钙质板岩
caldera	破火山口
caldera lake	火山口湖
Caledonian	加里东期
California	加利福尼亚州(美)
Californium(Cf)	锎
Calymmian	盖层系
Cambrian	寒武系
Cambrian explosion	寒武纪生物爆发
Cambrian Period(System)	寒武纪(系)
Cambridge blue	浅蓝
camel	驼色
Cameroon	喀麦隆
Canada	加拿大
Canada balsam	加拿大树胶
canary	淡黄
cancrinite	钙霞石
cannel coal	烛煤
cape	海岬,岬角
Cape Verde	佛得角
capillary aggregate	毛发状集合体
capillary concentration	毛细管浓缩
capillary water	毛细水
capping	矿帽
carboborite	水碳硼石
carbon cycle	碳循环
carbon nanotube	碳纳米管
Carbon(C)	碳

carbonaceous chondrite	碳质球粒陨石
carbonaceous siliceous-pelitic rock type U-ore	碳硅泥岩型铀矿
carbonate	碳酸盐
carbonate buildup	碳酸盐岩隆
carbonate compensation depth(CCD)	碳酸盐补偿深度
carbonate platform	碳酸盐台地
carbonate precipitation kinetics	碳酸盐沉淀动力学
carbonate rocks	碳酸盐岩
carbonate-apatite	碳磷灰石
carbonation	碳酸化
carbonatite	碳酸岩
carbonatite-type rare earth deposit	碳酸岩型稀土矿床
carbonatization	碳酸盐岩化
C-lattice	底心格子(C-格子)
Carboniferous	石炭系
Carboniferous Period(System)	石炭纪(系)
carbonado	黑金刚石
Carlin-type gold deposit	卡林型金矿床
Carlsbad twin	卡尔斯巴双晶,卡式双晶
Carlsbad-albite compound twin	卡-钠复合双晶
carmine	胭脂红,洋红,深红
carnallite	光卤石
carnallite deposit	光卤石矿床
carnegieite	三斜霞石
carpholite	纤锰柱石
carrying capacity	承载能力
cassiterite	锡石
cassiterite-quartz vein	锡石-石英脉
cassiterite-sulphide deposit	锡石-硫化物矿床
casts	铸模
catabolism	分解代谢作用
cataclasite	碎裂岩
cataclastic flow	碎裂流动

cataclastic texture	碎裂结构
catastrophism	灾变论
Cathaysia	华夏古大陆
cathodoluminescence	阴极发光
Cathodoluminescence Microscopy (CLM)	阴极发光显微镜
cation	阳离子
cave	溶洞,洞穴
cavity filling	充填[作用]
Cayman Islands	开曼群岛(英)
CBED	会聚束电子衍射
CCD	碳酸盐补偿深度
crustal low-velocity zone	地壳低速带
CDT standard(Canyon Diablo meteorite troilite standard)	CDT 标准(硫同位素标准,美国亚利桑那铁陨石中的陨硫铁的硫同位素比值)
Ce anomaly	铈异常
celandine green	浅灰绿色
celestite	天青石
celeste	天蓝
celestial body	天体
cell	晶胞
cell parameter	晶胞参数
cell sampling	格子采样
cellophane paper	透明纸
celsian	钡长石
cement	胶结物
cement generation	胶结物的世代
cement limestone	水泥石灰岩
cement raw materials	水泥原料
cementation	胶结,胶结作用
cementation type	胶结类型
Cenozoic (CZ)	新生代(界)
Cenozoic Era(Erathem)	新生代(界)

center of symmetry	对称中心
Central African Republic	中非共和国
central eruption	中心[式]喷发
Central-African copper belt	中非铜矿床带
CEPHPT	高压高温实验岩石学委员会
ceramic material	陶瓷原料矿产
ceramic raw materials	陶瓷原料
Cerium(Ce)	铈
cerulean	天蓝
cerussite	白铅矿
cervantite	黄锑矿
cesium deposit	铯矿床
CFT	晶体场理论
CGMW	世界地质图委员会
chabazite	菱沸石
Chad	乍得
chaidamuite	柴达木石
chain silicate	链状硅酸盐
chalcedonization	玉髓化
chalcedony	玉髓
chalchuite	翡翠
chalcocite	辉铜矿
chalcophile element	亲铜元素,亲硫元素
chalcopyrite	黄铜矿
chalk deposit	白垩矿床
chamosite	鲕绿泥石
channel bar	心滩
charcoal gray	炭灰色
charnockite	紫苏花岗岩
chatoyancy	猫眼,虎眼
chelate	螯合物
chemical and fertilizer raw materials	化工和化肥原料
chemical compaction	化学压实

chemical evolution in prebiological period 前生期化学演化
chemical evolution of environment 环境化学演化
chemical geodynamics 化学地球动力学
chemical kinetics 化学动力学
chemical petential 化学位
chemical sedimentary rock 化学沉积岩
chemical sedimentary structures 化学沉积构造
chemical species 化学物种
chemical weathering 化学风化
Chengjiang fauna 澄江动物群
chert 燧石
cherty iron formation 燧石铁建造
chevron fold 尖棱褶皱
CHGS 地质科学史委员会
chiastolite 空晶石
Chiatura manganese deposit 恰图拉矿床
Chile 智利
chill zone 冷凝带
chiluite 赤路矿
China 中国
chlorapatite 氯磷灰石
chlorargyrite 角银矿
Chlorine(Cl) 氯
chlorite 绿泥石
chloritization 绿泥石化
chloritoid 硬绿泥石
chlorophyllinite 叶绿素体
chocolate 巧克力色,红褐色,赭[石]色
chondrite 球粒陨石
chondrodite 粒硅镁石
chondrule [陨石]球粒
chromate 铬酸盐
chrome-diopside 铬透辉石

chrome-spinel	铬尖晶石
chromite	铬铁矿
chromite deposit of Alpine-type	阿尔卑斯型铬铁矿床
chromite ore magma	铬铁矿浆
chromium deposit	铬矿床
Chromium(Cr)	铬
chromophores	色素离子
chron	时
chronostratigraphic unit	年代地层单位
chronozone	时带
chrysocolla	硅孔雀石
chrysolite	贵橄榄石
chrysotile	纤蛇纹石,温石棉,蛇纹石石棉
chrysotile-asbestos deposit	蛇纹石石棉矿床
CHUR(chondritic uniform reservoir)	CHUR 值(球粒陨石型均一岩浆房的地幔库)
Ciel	天蓝
CIG	国际地球物理学委员会
cinder	火山渣
cinder cone	火山渣锥
cinnabar	辰砂
CIPASH	大气科学和水文学国际规划委员会
circum-Pasific metallogenic belt	环太平洋成矿带
Cisuralian	乌拉尔统
citreous	柠檬[黄]色
citrine	黄水晶,柠檬[黄]色
clarain	亮煤
claret-colored	深紫红色的
clarite	微亮煤
Clarke value	克拉克值,地壳元素丰度
Clarke	克拉克值
Clarke F W	克拉克

Clarke of atom	原子克拉克值
Clarke of concentration	浓度克拉克值
class	大类(矿物分类单位)
class of symmetry	对称型
clast	碎屑
clay deposit	粘土矿床
clay mineral	粘土矿物
clayization	粘土化
cleavage	解理,劈理
cleavage domain	劈理域
cleavage fan	劈理扇
cleavage in trace	极不完全解理
cleavage plane	解理面
cleavelandite	叶钠长石
Clerici's solution	克列里奇液
climate	气候
climate system	气候系统
climatic belt	气候带
climatic type	气候型
climatic zone	气候带
climax	顶级群落
clinochlore	斜绿泥石
clinoenstatite	斜顽辉石
clinoform	斜坡地形,倾斜型,斜坡沉积
clinoptilolite	斜发沸石
clinopyroxene	单斜辉石
clinopyroxenite	单斜辉石岩
clinozoisite	斜黝帘石
CLM	阴极发光显微镜
closed form	闭形
closed system	封闭系统
closer packed	紧密堆积
closure	闭合度

closure temperature	封闭温度
clot	凝块,团块,斑块,凝块石
cloud	云
clue for prespecting	找矿标志
cluster of galaxies	星系团
CMAFM	接触模式原子力显微镜
CMG	海洋地质学委员会
CN	配位数
CNMMNIMA	国际矿物学会新矿物和矿物命名委员会
coal	煤[炭]
coal accumulating region	聚煤区
coal accumulation	聚煤作用
coal ball	煤核
coal basin	煤盆地,聚煤盆地
coal diagenesis	煤成岩作用
coal facies	煤相
coal gas	煤成气
coal geochemistry	煤地球化学
coal geology	煤[田]地质学
coal measures	含煤岩系(煤系)
coal metamorphism	煤变质作用
coal mining geology	煤矿地质
coal petrography	煤岩学
coal petrology	煤岩学
coal rank	煤级
coal seam	煤层
coal-bearing series	含煤岩系(煤系)
coal-black	墨黑的
coalfield	煤田
coal-forming material	成煤物质
coal-forming period	聚煤期
coalification	煤化作用

coarse	粗粒的，粗糙的
coarse grained	粗粒的
coarse grained texture	粗粒结构
coarse sandstone	粗砂岩
coarse wacke	粗粒杂砂岩
coarse-grained	粗粒的
coarse-tail grading	粗尾粒序
coast	海岸
cobalt blue	钴蓝色，艳蓝色
Cobalt(Co)	钴
cobaltite	辉砷钴矿
cobble conglomerate	粗砾岩
coccolith	颗形石，颗石，颗石藻，球石
coccolithophora	圆锥颗石，颗石鞭毛藻
coefficient	系数，常数
coesite	柯石英
coffee	咖啡色，深褐色
coffinite	铀石
coking coal	焦煤
cold extraction	冷提取
cold seal pressure vessel	冷封口高压容器
collapse earthquake	陷落地震
collinite	无结构镜质体
collision-type orogeny	碰撞型造山[作用]
collocrystallization	胶体结晶作用
colloform structure	胶状构造
colloid	胶体，胶粒
colloidal deposition	胶体沉积
colloidal mineral	胶体矿物
cellophane	胶磷矿
Colombia	哥伦比亚
color	颜色
color centre	色心

color index	颜色指数(色率)
color of mineral	矿物颜色
Colorado	科罗拉多州(美)
colorful	鲜艳的
colorless	无色的
columnar	柱状
columnar aggregate	柱状集合体
columnar jointing	柱状节理
comb structure	梳状构造
combination drive reservoir	混合驱动储油层
combination form	聚形
combination striations	聚形纹
combustibility	可燃性
comet	彗星
commercial type of mineral deposits	矿床工业类型
Commission for Marine Geology (CMG)	海洋地质学委员会
Commission for the Geological Map of the World (CGMW)	世界地质图委员会
Commission on Experimental Petrology at High Pressures and Temperatures (CEPHPT)	高压高温实验岩石学委员会
Commission on New Minerals and Mineral Names of the International Mineralogical Association (CNMMNIMA)	国际矿物学会新矿物和矿物命名委员会
Commission on Recent Crustal Movement (CRCM)	近代地壳运动委员会
Commission on Systematics in Petrology (CSP)	岩石分类学委员会
Commission on Tectonics (CT)	大地构造学委员会
Committee on History of Geological Sciences (CHGS)	地质科学史委员会
Committee on International Geophysics (CIG)	国际地球物理学委员会
Committee on International Programs in Atmospheric Sciences and Hydrology (CIPASH)	大气科学和水文学国际规划委员会

Committee on Science and Technology in Developing Countries (CSTDC)	发展中国家科学技术委员会
common lead	普通铅
common net	共格面网
Comoros	科摩罗
compaction	压实,压缩,压紧
comparative crystal chemistry	比较晶体化学
comparative planetology	比较行星学
compass	罗盘
compatible element	相容元素
compensator	补色器,消色器
complete extinction	全消光
complete isomorphous series	完全类质同象
complex	杂岩
composite volcano	复合火山
composition plane	双晶接合面
composition surface	双晶接合面
compositional layering	成分层
compositional maturity	成分成熟度
composition-paragenesis diagram	成分-共生图解
compound ooid	复鲕
compound twin	复合双晶
compound twinning	复合双晶
compound volcano	复合火山
compressibility	可压缩性
Computed X-ray Tomography (CXT)	程序控制 X 射线层析
computer model	计算机模拟
concealed ore body	隐伏矿体
concentration	富集作用
concentration center	浓集中心
concentration coefficient	浓集系数,富集系数
concentration gradient	浓度梯度
concentration of element	元素浓度

concentric ooid	同心鲕
concentric structure	同心环状构造
conchoidal fracture	贝壳状断口
concordant intrusion	整合侵入体
concordia age	一致年龄
concretion	结核体
condensate field	凝析气田
condensate oil	凝析油
condensation	凝结,凝聚
condensation of elements	元素凝聚
conditional reserve	条件储量
conduction	传导
conduction band	导带
conductive heat flow modeling	传导热流模拟
conductivity	传导性,导电性
conduit	火山通道
conflagration	火灾
Confocal Scanning Optical Microscopy (CSOM)	共焦扫描光学显微镜
conformity	整合
conglomerate	砾岩
conglomerate type U-ore	砾岩型铀矿
Congo	刚果
conical fold	锥状褶皱
conjugate joint	共轭节理
Connecticut	康涅狄格州(美)
Conrad discontinuity	康拉德不连续面
constant form	定形
constant zone of subsurface temperature	常温层
constitution water	结构水
construction	作图
constructional materials	建筑材料
constructional terrace	堆积阶地

consumer	消费者
contact aureole	接触晕
contact cementation	接触式胶结
contact metamorphic rock	接触变质岩
contact metasomatic rock	接触交代岩
contact thermometamorphic rock	接触热变质岩
contact twin	接触双晶
Contacting Mode Atomic Force Microscopy (CMAFM)	接触模式原子力显微镜
contact-metasomatic deposit	接触交代矿床
contact-metasomatic metamorphism	接触-交代变质[作用]
contact-thermal metamorphism	接触-热变质[作用]
contamination	玷污
continent	大陆
continental drift	大陆漂移
continental drift theory	大陆漂移说
continental glacier	大陆[性]冰川
continental heat flow	大陆热流
continental rift system	大陆裂谷系
continent-continent collision	陆-陆碰撞
continuity	连续
continuous cleavage	连续劈理
continuous reaction	连续反应
continuous transformation	连续相变
contraction tectonics	收缩构造
contraction theory	收缩说
contrast	衬度
convection	对流[作用]
convective circulation	对流循环
convective hydrothermal cell	水热对流单元
convergence	幅合
convergent	会聚光
convergent light figure	干涉图

convergent plate boundary	聚敛板块边界,汇聚型板块边界
convergent plate margin	汇聚型板块边缘
Convergent-Beam Electron Diffraction (CBED)	会聚束电子衍射
converse piezoelectricity	电致伸缩(反压电性)
convolute bedding	包卷层理,旋绕层理
Cook Islands	库克群岛(新)
coordinate direction	坐标方向
coordination number(CN)	配位数
copper	自然铜
copper deposit	铜矿床
Copper(Cu)	铜
copper-nickel deposit	铜镍矿床
copper-nickel sulphide deposit	铜镍硫化物矿床
coquina	贝壳灰岩
coral	珊瑚，珊瑚虫
coral reef	珊瑚礁
cordierite	堇青石
core	核部
Coriolis force	科里奥利力
corona	冠状体
correlation of coal seam	煤层对比
corroded funnel	溶斗
corundum	刚玉
cosmic abundance	宇宙丰度
cosmic dust	宇宙尘
cosmic geology	宇宙地质学
cosmic mineralogy	宇宙矿物学
cosmic ray exposure age	宇宙线暴露年龄
cosmic rays	宇宙射线
cosmic spherule	宇宙颗粒
cosmochemistry	宇宙化学
cosmochronology	宇宙年代学

cosmogenic hypothesis of mineralization	宇宙源成矿说
cosmogenic nuclide	宇宙成因核素
cosmogonic theory	天体演化理论,宇宙起源说
cosmos	宇宙
Costa Rica	哥斯达黎加
Cote d'Ivoire	科特迪瓦
cotectic crystallization	同结晶,共结晶
cotype	共型
country rock	围岩
country rock alteration	围岩蚀变
coupled reaction	耦合反应
covalency	共价
covellite	铜蓝
cover slip	盖玻片
cover strata	盖层
cpl	正交偏光
crater	火山口,环形山,陨击坑
crater lake	火山口湖
craton	稳定地块，克拉通
CRCM	近代地壳运动委员会
creep behaviour	蠕变性
crenulation lineation	皱纹线理
crescent lake	月牙湖
crest line	脊线
Cretaceous	白垩系
Cretaceous Period(System)	白垩纪(系)
crevice water	裂隙水
crimson	赤红,深红,绛色
cristobalite	方石英
critical concentration	临界浓度
critical mineral assemblage	临界矿物组合
critical radius	临界半径
critical resource	临界资源

Croatia	克罗地亚
croci	橘黄色
crocidolite deposit	蓝石棉矿床，青石棉矿床
crocidolite-asbestos deposit	蓝石棉矿床
crocoite	铬铅矿
cross fold	横褶皱
cross section	横剖面
cross-bedding	交错层理
crosshatch twinning	格子状双晶
crossite	青铝闪石
cross-lamination	交错纹层
Crossopterygii	总鳍鱼类
cross-polarized light(cpl)	正交偏光
crude oil	原油
crumpled texture	揉皱结构
crush breccia	压碎角砾岩
crush conglomerate	压碎砾岩
crush texture	压碎结构
crust	地壳
crustal movement	地壳运动
crust-derived	壳源[的]
crustified vein	皮壳状脉
crusty structure	皮壳状构造
Cryogenian	成冰系
cryolite	冰晶石
cryopedological action	冻融作用
cryopedology	冻土学
cryptocrystalline	隐晶质
cryptocrystalline aggregate	隐晶集合体
cryptoperthite	超显微条纹长石(电子显微镜下)
Cryptozoic Eon(Eonothem)	隐生宙(宇)
crystal	晶体(结晶体)
crystal casts	假晶

crystal category	晶族
crystal class	晶类
crystal constants	晶体常数
crystal defect	晶体缺陷
crystal edge	晶棱
crystal face	晶面
crystal field theory(CFT)	晶体场理论
crystal form	晶形
crystal growth	晶体生长
crystal habit	结晶习性
crystal indices	结晶符号,晶体符号
crystal nucleus	晶芽,晶核
crystal optics	晶体光学
crystal orientation	晶体定向
crystal physics	晶体物理
crystal pyroclast	[火山]晶屑
crystal system	晶系
crystal zone	晶带
crystalline dolomite	结晶白云岩
crystalline limestone	结晶灰岩
crystalline mimetic	拟晶
crystalline rock	结晶岩
crystalline substance	[结]晶质
crystalline schist	结晶片岩
crystalline texture	结晶结构
crystallinity	结晶度
crystallite	雏晶
crystallization	结晶作用
crystallization index	结晶指数
crystallization water	结晶水
crystallization-differentiation deposit	结晶分异矿床
crystallizing	晶化
crystalloblastic series	变晶系

crystallochemical formula	晶体化学式
crystallochemistry	晶体化学
crystallographic axis	结晶轴
crystallographic form	结晶单形
crystallography	结晶学
crystallology	晶体结构学
CSOM	共焦扫描光学显微镜
CSP	岩石分类学委员会
CSTDC	发展中国家科学技术委员会
CT	大地构造学委员会
ctenoid texture	栉壳状结构
Cuba	古巴
cube	立方体
cubic system	立方晶系,等轴晶系
cubic closest packing	立方最紧密堆积
culmination	高点,褶隆
cummingtonite	镁铁闪石
cumulate	堆晶岩
cupola hypothesis of mineralization	岩钟成矿说
cupreous sandstone	含铜砂岩
cuprite	赤铜矿
Curie temperature	居里温度
Curie theory	居里原理
Curie-Wulff theory	居里-吴里弗原理
Curium(Cm)	锔
current ripple	流痕
curvature	曲率
cuspate-lobate folds	尖-圆形褶皱
cuticolor	肉色
cutinite	角质体
cutoff grade	边界品位
CXT	程序控制 X 射线层析
cyan	青色,蓝绿色

cyaneous	深蓝色
cyanobacteria	蓝藻细菌
cycles	旋回
cyclic reaction	循环反应
cyclic twin	轮式双晶(环状双晶)
cyclone	气旋
cyclosilicate	环状硅酸盐
cylindrical and helicitical	卷曲结构
cylindrical fold	圆柱状褶皱
Cyprus	塞浦路斯
Czech Rep.	捷克

D

DAC	钻石高压盒
dacite	英安岩
damask	淡红色
danbaite	丹巴矿
Danny-Harker rule	唐奈-哈克原理
daomanite	道马矿
daqingshanite	大青山矿
dark	暗色,深色的
dark blue	暗蓝色,深蓝色
dark brown	茶[褐]色,棕褐色,咖啡色,深褐色
dark green	深绿
dark grayish brown	深灰褐色
dark-colored mineral	暗色矿物
darkish	浅黑的
data quality monitoring	质量监控
datolite	硅硼钙石
Dauphine twin	多菲内双晶(曾用名:道芬双晶)
DD	密度测试
debris	岩屑,碎石,碎屑,碎石堆,有机物残渣
debris flow	泥石流,碎屑流

Debye-Scherrer method	德拜-谢勒法
decarbonation reaction	脱碳酸反应
decolorization	褪色作用
decomposer	分解者,分解体
decrepitation method	爆裂法
dedolomitization	去白云岩化,退白云岩化
deep burial diagenesis	深埋成岩作用
deep red	绛色,深红色
deep-focus earthquake	深源地震
deep-sea	深海
deep-sea fan	深海扇
deerite	迪尔石
deflation	吹蚀作用
deflation unaka	风蚀残丘
deformation	变形
deformation band	变形带
deformation boulder	变形砾石
deformation lamellae	变形纹
deformation texture	变形结构
deformation twins	变形双晶
deformational sedimentary structures	变形沉积构造
degasification	去气作用
degenerated system	简并系统
degradation	降解,陵夷[作用]
degradation	降解作用
dehydration	脱水作用
dehydration experiment	脱水实验
dehydration reaction	脱水反应
dehydration-reaction isograd	脱水反应等变线
delamination	拆沉
Delaware	特拉华州(美)
delta	三角洲

delta 7/4 ＝($^{207}Pb/^{204}Pb-^{207}Pb/^{204}PbNHRL$) * 100, $^{207}Pb/^{204}PbNHRL = 15.627 + 1.209$ ($^{207}Pb/^{204}Pb$)	铅 207 富集指数
delta 8/4 ＝($^{208}Pb/^{204}Pb-^{208}Pb/^{204}PbNHRL$) * 100, $^{208}Pb/^{204}PbNHRL = 13.491 + 0.1804$ ($^{208}Pb/^{204}Pb$)	铅 208 富集指数
delta Sr	锶富集指数
delta value	delta 值[绝对变差 δ,反映同位素组成与标准相比重同位素的富集或亏损程度(‰)]
deltoid	偏方形
deluvial placer	坡积砂矿
demorphism	同质二象
dendritic aggregate	树枝状集合体
Denmark	丹麦
density	密度
density compensation	密度补偿
density current	密度流
density determination (DD)	密度测试
denudation	剥蚀[作用]
depleted mantle(DM)	亏损地幔
depleted MORB mantle(DMM)	亏损的 MORB 地幔
depletion drive reservoir	溶解气驱储油层
depletion of element	元素亏损
deposition	沉积[作用],沉积物
depositional process	沉积作用
depositional setting	沉积建造
depth compensation	深度补偿
depth zone	深度带
descending solution	下降溶液
descloizite	羟钒锌铅石
desert area	荒漠区,沙漠区
desertification	沙漠化,荒漠化
deserts	沙漠

determinative mineralogy	鉴定矿物学
deltohedron	偏十二面体
detrital	碎屑的
detrital texture	碎屑结构
development line	演化线
development well	开发井
deviatoric stress	偏应力
devitrification	脱玻化,晶化
devolatilization reaction	脱挥发分反应
devolution	崩塌
Devonian Period(System)	泥盆纪(系)
dew	露
dewatering	脱水
DI	分异指数
diabase	辉绿岩
diabasic texture	辉绿结构
diablastic texture	横交结构
diachronous	穿时的
diachronous-progressive metamorphism	穿时递增变质[作用]
diadochy	互代性,晶格同位
diagenesis	[沉积]成岩作用,岩化作用
diagenetic	成岩作用的
diagenetic stages	成岩阶段
diagram of element ratio	元素比值图
diakisdodecahedron	偏方复十二面体
diamagnetism	抗磁性,逆磁性,反磁性
diamictite	混杂陆源沉积岩,混积岩
diamond	金刚石,钻石
diamond anvil cell	金刚石压腔
Diamond Anvil Cell (DAC)	钻石高压盒
diamond deposit	金刚石矿床
Diaoyudaoite	钓鱼岛石
diaphaneity	透明性

diaphthorite	退变岩
diapir structure	底辟构造
diaplectic glass	击变玻璃
diaspore	硬水铝石
diatexite	全熔岩
diatom	硅藻类
diatomite	硅藻岩
diatomite deposit	硅藻土矿床
diatreme	火山管
DIC	微分干涉相衬显微镜
dichroism	二[向]色性
dickite	迪开石
decussate texture	交叉结构
dielectric constant	介电常数,电容率
dielectricity	介电性
Differential Interference Contrast Microscopy (DIC)	微分干涉相衬显微镜
Differential Laser Interferometry(DLI)	微分激光干涉仪
differential mobility of components	组分差异活动性
differential rotation of earth core	地核差异旋转
Differential Scanning Calorimeter (DSC)	差式扫描量热分析
Differential Thermal Analysis (DAT)	差热分析
differential weathering	差异风化
Differentiation Index(DI)	分异指数
Diffuse Reflectance Infrared Fourier Transfer Spectroscopy (DRIFT)	漫反射红外傅立叶转换光谱
diffusion	扩散[作用]
diffusion halo	扩散晕
diffusional fractionation	扩散分馏
diffusive flow	扩散流动
diffusive metasomatism	扩散交代[作用]
digital earth	数字地球
dihedral angle	二面角

dihexagonal bipyramid	复六方双锥
dihexagonal prism	复六方柱
dihexagonal pyramid	复六方单锥
dike	岩墙
dimension stone	石材
dimensional preferred orientation	形态优选方位
dimetasomatism	双交代作用
dimmer	遮光器
dingy	暗黑色
diopside	透辉石
diorite	闪长岩
dip	倾向
dip angle	倾角
dip isogon	等倾斜(角)线
diplogenetic deposit	二重成因矿床
dipping structure	倾斜构造
dip-slip-fault	倾向滑动断层
dipyramid	双锥
direct lattice	正格子(与倒易格子对应)
directed pressure	定向压力
direction of extinction	消光位(消光方向)
directional structure	定向构造
dirt band	夹矸
disconformity	假整合,平行不整合
discontinuity	不连续
discontinuous reaction	不连续反应
discontinuous transformation	不连续相变
discordant intrusion	不整合侵入体
discordant age	不一致年龄
disequilibrium texture	不平衡结构
disharmonic fold	不谐调褶皱
disilicate	双硅酸盐
disintegration	崩解,剥蚀,分解,蜕变

dislocation	位错
dislocation breccia	断层角砾岩
dislocation metamorphic rock	断层变质岩
disorder structure	无序结构
dispersed element	分散元素
dispersion	分散,色散
dispersion coefficient	分散系数
dispersion field	分散场
dispersion halo	分散晕
dispersion of Becke line	贝克线色散
dispersion of element	元素分散
dispersion of optic angle	光轴角色散
dispersion of refringence	双折射率色散
dispersion train	分散流
displacement	位移
displacive transformation	位移型转变,位移式相变
disseminated structure	浸染状构造
dissolution	溶解,溶解作用
dissolution basin	溶蚀盆地
dissolution kinetics	溶解动力学
dissolved gas	溶解气
dissolvent	溶剂
dissociation	分解
distal ore deposit	远火山活动金属矿床
disthene	蓝晶石
distorted crystal	歪晶
distortion	形变,畸变,变形
distribution coefficient	分配系数
distribution diagram of element	元素分布图
distribution grading	配分粒序
distribution of element	元素分布
distribution of n-alkanes	正烷烃分布
distribution pattern of rare earth element	稀土元素分布模式

District of Columbia	哥伦比亚特区(美)
ditetragonal bipyramid	复四方单锥
ditetragonal prism	复四方柱
ditetragonal pyramid	复四方单锥
ditetragonal scalenohedron	复四方偏三角面体
ditrigonal bipyramid	复三方双锥
ditrigonal prism	复三方柱
ditrigonal pyramid	复三方单锥
ditrigonal scalenohedron	复三方偏三角面体
divariant equilibrium	双变平衡
divariant field	双变区
divergence	幅散
divergent plate boundary	离散[型]板块边缘
Djibouti	吉布提
DLI	微分激光干涉仪
DM	亏损地幔
DMM	亏损的 MORB 地幔
dodecahedron	菱形十二面体
dolerite	粗玄岩,辉绿岩
dolomite	白云岩,白云石
dolomite deposit	白云岩矿床
dolomitite	白云岩
dolomitization	白云石化[作用]
dolostone	白云岩
domain	晶畴
domain structure	畴结构
dome	穹隆,坡面
Doppler effect	多普勒效应
dorag dolostone	混合水白云岩,混合白云石化作用
dormant volcano	休眠火山
double chain silicate	双链状硅酸盐
downcurrent	顺流
down-cutting	下切[作用]

downthrown side	下降盘
drainage anomaly	水系异常
draping structure	披覆构造
dravite	镁电气石
DRIFT	漫反射红外傅立叶转换光谱
drought	干旱
drumlin	鼓丘
druse	晶簇
drusy structure	晶簇构造
dry gas	干气
dry valley	干谷
DSC	差式扫描量热分析
ductility	延[展]性
dull	暗色
dull coal	暗淡煤
dull grayish brown	暗灰棕色
dumortierite	蓝线石
dun	暗褐色
dunes	沙丘
dunite	纯橄榄岩
dunn bass	夹石
DUPAL(Hart (1984) noticed that oceanic basalts with high $^{207}Pb/^{204}Pb$ and $^{208}Pb/^{204}Pb$ ratios for a given value of $^{206}Pb/^{204}Pb$ come mainly from a belt centered at about 30°S. Hart named this feature the DUPAL anomaly (after Dupré and Allègre). He defined the DUPAL isotopic signature as having higherΔSr($\Delta Sr = [^{87}Sr/^{86}Sr - 0.7030] \times 104$) and high $\Delta 8/4$ and $\Delta 7/4$)	DUPAL异常(Hart (1984)注意到具有$^{207}Pb/^{204}Pb$和$^{208}Pb/^{204}Pb$对给定的$^{206}Pb/^{204}Pb$值的高比值的大洋玄武岩主要来自集中于约30°S的带。Hart将这一特征命名为DUPAL异常(以Dupré和Allègre命名)。他将该DUPAL同位素识别标志定义为较高的ΔSr($\Delta Sr=[^{87}Sr/^{86}Sr-0.7030] \times 104$)以及高$\Delta 8/4$和$\Delta 7/4$)
durain	暗煤

durite	微暗煤
dust	漂尘
dust bowl	尘暴
dust devil	尘暴
dust storm	尘暴
dustfall	降尘
dyke	岩墙
dynamic metamorphism	动力变质[作用]
dynamic recrystallization	动态重结晶
dynamics	动力学
dynamics of resource concept	资源概念的动态性
dynamometamorphic rock	动力变质岩
dynamothermal metamorphic rock	动热变质岩
Dysprosium(Dy)	镝
Dzhezkazgan copper deposit	哲兹卡兹甘铜矿床

E

early diagenesis	早期成岩作用
early magmatic deposit	早期岩浆矿床
early magmatic segregation-type deposit	早期岩浆分凝型矿床
Earth	地球
earth axis	地轴
earth core	地核
earth crust	地壳
earth magnetic field	电磁场
earth material science	地球物质科学
earth oscillation	地球震荡
earth pole	地极
earth science	地球科学,地学
earth slide	滑坡
earth surface science	地球表层科学
earth system	地球系统
earth system science	地球系统科学

earth tide	固体潮
earth-like planet	类地行星
earthquake	地震
earthquake effect	地震效应
earthquake intensity map	地震烈度图
earthquake prediction	地震预报
earthy aggregate	土状光泽
earthy luster	土状集合体
East Timor	东帝汶
Eastern Samoa	东萨摩亚(美)
eastonite	镁叶云母
eccentric ooid	偏心鲕
eclogite	榴辉岩
eclogite facies	榴辉岩相
ecological balance	生态系统平衡
ecological factor	生态因素
ecology	生态学
economic geology	经济地质学
economic mineral deposit	经济矿床
ecosystem	生物系[统]
ecstatic sea level	理论海平面
ectasian	延展系
Ecuador	厄瓜多尔
ED	电子衍射
edenite	浅闪石
EDS	能量分散光谱
EDX	能量分散 X 射线光谱
EDXS	能量分散 X 射线光谱
EELS	电子能损光谱
effective permeability	有效渗透率
Egypt	埃及
EH(oxidation-redution potential)	氧化还原电位
eigenvector	本征矢量,特征向量

Einsteinium(Es)	锿
El Nino	厄尔尼诺现象
El Salvador	萨尔瓦多
elastic behavior	弹性行为
elasticity	弹性
elbaite	锂电气石
elbow twin	肘状双晶
electrical conductivity	导电性
Electro Microprobe Analysis (EMPA)	电子探针分析
electro negativity	电负性
electroluminescence	场致发光
electromagnetism	电磁性
electron affinity	电子亲和性
electron configuration	电子构型
Electron Diffraction (ED)	电子衍射
Electron Energy-Loss Spectroscopy (EELS)	电子能损光谱
Electron Microprobe Analysis-Mossbauer Spectroscopy (EMPA/Mossbauer)	电子探针穆斯保尔光谱
Electron Paramagnetic Resonance Spectrometry (EPR, EPRS)	电子顺磁共振光谱
Electron Probe Microanalyzer (EPMA)	电子探针分析
Electron Spin Resonance Spectroscopy (ERS, ERS)	电子自旋共振光谱
electronegativity	电负性
electrostriction	电致伸缩(反压电性)
electrum	银金矿
element	单质,元素
element association	元素组合
element pair	元素对
element ratio	元素比值
element reactivation	元素活化
element substitution	元素置换

element transfer	元素迁移
element's abundance	元素丰度
elements of earthquake prediction	地震预报要素
elements of fold	褶皱要素
elliptical polarization	椭圆偏振
elongation	延长
elongation sign	延长符号
eluvial placer	残积砂矿
eluvium	残积物
EM Ⅰ	富集地幔Ⅰ
EM Ⅱ	富集地幔Ⅱ
embayed texture	港湾结构
embayment	海湾,涨湾,支地槽,海湾状熔蚀
emerald	祖母绿
emerald	翠绿
emerald green	鲜绿色
emfola twin	恩福拉双晶
eminent cleavage	极完全解理
Emission Spectroscopy (ES)	发射光谱
EMPA	电子探针分析
EMPA/Mossbauer	电子探针穆斯保尔光谱
emulsion texture	乳滴状结构
enantiomorphic relationship	左右形关系
enantiotropic transformation	可逆的(双向的)转变
enclave	包体
enclosure	包体
encrinite	海百合,海百合灰岩
end moraine bar	终碛堤
endemic disease	地方病
endogenesis	内生作用
endogenetic anomaly	内生异常
endogenetic force	内营力,内动力
endogenic mineral deposit	内生矿床

endogenic process	内生作用
endonenic metallization	内生成矿作用
Energy-Dispersive Spectroscopy (EDS)	能[量色]散谱仪
Energy-Dispersive X-ray Spectroscopy (EDX, EDXS)	能量分散 X 射线光谱
enlargement texture	加大边结构
enriched mantle Ⅰ(EM Ⅰ)	富集地幔Ⅰ
enriched mantle Ⅱ(EM Ⅱ)	富集地幔Ⅱ
enrichment of element	元素富集
enstatite	顽火辉石
enstatite chondrite	顽辉石球粒陨石
enthalpy (H)	焓,热函
entropy (S)	熵
enveloping surface	包络面
environment protection	环境保护
environmental anomaly	环境异常
environmental background	环境背景[值]
environmental chronology	环境年代学
environmental degradation	环境退化
environmental effect	环境效应
environmental factor	环境因子
environmental geochemistry	环境地球化学
environmental geochemistry parameter	环境地球化学参数
environmental impact assessment	环境影响评价
environmental influence	环境影响
environmental interface	环境界面
environmental interface geochemistry	环境界面地球化学
environmental medium	环境介质
environmental organic geochemistry	环境有机地球化学
environmental planning	环境规划
environmental pollution	环境污染
equidimensional	等轴的
even-grained	等粒结构

exsolution	出熔[作用]
environmental quality	环境质量
environmental quality assessment	环境质量评价
environmental quality atlas	环境质量图
environmental quality index	环境质量指数
environmental quality parameter	环境质量参数
environmental quality standard	环境质量标准
environmental quality variation	环境质量变异
environmental regionalization	环境区划
environmental reservoir	环境储库
environmental self-purification	环境自净
environmental simulation	环境模拟
environmental transition	环境变迁
Eoarchean	始太古界
Eocene	始新统
Eocene Epoch(Series)	始新世(统)
Eon	宙
Eonothem	宇
epeiric sea	陆表海,边缘海,浅海
epeirogenetic movement	造陆运动
epeirogeny	造陆作用
epibolite	顺层混合岩
epicenter	震中
epicentral intensity	震中烈度
epiclast	表生碎屑,外来碎屑
epidote	绿帘石
epidote amphibolite	绿帘角闪岩
epidote-amphibolite facies	绿帘-角闪岩相
epidotite	绿帘石岩
epidotization	绿帘石化
epigenetic anomaly	后生异常
epigenetic deposit	后生矿床
epimorphism	后继形貌

episode	幕(期)
epitaxial	外延的,定向附生的
epitaxial growth	浮生(外延生长)
epitaxic	外延的,定向附生的
epithermal deposit	低温水热矿床,低温热液矿床,浅成热液矿床
epizone	浅带
EPMA	电子探针分析
epoch	世
epoch of mineralization	成矿期
EPR	电子顺磁共振光谱
EPRS	电子顺磁共振光谱
epsomite	泻利盐
Equatorial Guinea	赤道几内亚
equidimensional	等轴的
equigranular	等粒状
equilibrium constant	平衡常数
equilibrium domain	平衡畴
equilibrium phase diagram	平衡相图
equipoints	等效点系
era	代
erathem	界
Erbium(Er)	铒
Eritrea	厄立特里亚
erlianite	二连石
erosion	侵蚀[作用],溶蚀
erosion base level	侵蚀基面
erosional terrace	侵蚀阶地
ERS	电子自旋共振光谱
ertixiite	额尔齐斯石
eruption column	喷发柱
ES	发射光谱
esker	蛇丘

essexite	碱性辉长岩
Esterel twin	埃斯特勒双晶
Estonia	爱沙尼亚
etch figure	蚀像
etch pit	蚀坑
Ethiopia	埃塞俄比亚
Eu anomaly	铕异常
eucaryote	真核生物
eucrite	钙长辉长无球粒陨石,钙长辉长岩
European Associantion of Exploration Geophysicists (EAEG)	欧洲勘探地球物理学家协会
European Association of Earth Science Editors (Editerra) (EAESE)	欧洲地学编辑者协会
European Geophysical Society (EGS)	欧洲地球物理学会
Europium(Eu)	铕
eustacy	全球海平面变化
eustatic	海平面变化的
eutrophication	富营养化,超营养作用
euxinic	静海相的,闭塞环境的
evaluation well	评价井(详探井)
evaporate	蒸发,蒸发盐,蒸发岩
evaporite deposit	蒸发岩矿床(蒸发沉积矿床)
even fracture	平坦状断口
even rule	偶数规则
even-grained	等粒结构
even-odd mass effect	偶奇效应
Evler pole	欧拉极
evolution	演化
evolution theory	进化论
EXAFS	扩展 X 射线吸收精细结构光谱
excess argon	过剩氩
excess components	过剩组分
exchange components	交换组分

exchange equilibrium	交换平衡
exchange reaction	交换反应
EXELFS	扩展 X 射线能失精细结构光谱
exfoliation	层裂,剥离
exfoliation dome	剥离丘
exhalation deposit	喷气矿床,喷流矿床
exinite	壳质组,稳定组
existing form element	元素存在形式
exogenetic metallization	外生成矿作用
exogenesis	外生作用
exogenetic force	外营力,外动力
exogenic mineral deposit	外生矿床
exogenic process	外生作用
expanded perlite material deposit	膨胀珍珠岩原料矿床
expansion theory	膨胀说
experimental geochemistry	实验地球化学
experimental metallogeny	实验矿床学
experimental mineralogy	实验矿物学
experimental petrology	实验岩石学
exploration and prospecting geology	勘查地质学
exploration geochemistry	勘查地球化学
exploratory well	探井
explosive eruption	爆炸性喷发
explosive index	爆发指数
exsolution	固溶体离溶,固溶体分解,出溶[作用]
exsolution lamella	出溶条纹
exsolution texture	出溶结构,离溶结构
exsolved	出溶的
Extended X-ray Absorption Fine-Structure Spectroscopy (EXAFS)	扩展 X 射线吸收精细结构光谱
extension	伸展[作用]
external buffering	外缓冲

external geomorphic agency	地貌外营力
externally buffered components	外缓冲组分
extinct volcano	死火山
extinction	绝灭,消光
extinction angle	消光角
extinction direction	消光位
extraclast	外碎屑,外来碎屑
extraordinary ray	非[寻]常光
extraterrestrial civilization	地外文明
extraterrestrial metamorphism	地外变质[作用]
extrusive facies	喷出相
extrusive rock	喷出岩

F

fabric	构造，组构
face-centered lattice	面心格子(F-格子)
facies of ore body root	矿根相
facies of ore body top	矿顶相
facies sequences	相序
facing down fold	翻转褶皱
factor of soil formation	成土因素
Faeroe Islands	法罗群岛(丹)
fair cleavage	中等解理
fallout	散落物
false anomaly	假异常
false isograd	假等变线
famine	饥荒
fan fold	扇形褶皱
fanglomerate	扇积砾，扇砾岩
fat coal	肥煤
fault	断层
fault array	断裂阵列
fault breccia	断层角砾岩

fault clay	断层泥
fault earthquake	断层地震
fault geometry	断层几何学
fault gouge	断层泥
fault rock	断层岩
fault surface	断层面
fault trace	断层线
fault trap	断层圈闭
fault wall	断盘
fault zone	断层(裂)带
fawn	鹿毛色,浅黄褐色
fawn-coloured	淡黄褐色的
fayalite	铁橄榄石
FC	分离结晶作用
fecal pellet	粪团粒
Fedorov stage	弗氏旋转台
feldspar	长石
feldspar deposit	长石矿床
feldspar lithic sandstone	长石岩屑砂岩
feldspar sandstone	长石砂岩
feldspathic quartz sandstone	长石石英砂岩
feldspathoids	似长石
fels	粒岩
fels or rock	粒岩或××岩
felsic	长英质
felsic hornfels	长英角岩
felsic rock	长英质岩石
felsitic texture	霏细结构
fenster	构造窗
Fermium(Fm)	镄
ferrimagnetism	亚铁磁性
ferriphengite	铁硅白云母
ferromagnesian index	铁镁指数

ferromagnetism	铁磁性
ferrosilite	铁辉石
ferrous metal deposit	黑色金属矿床
ferrugineous	铁锈色的
ferruginous quartzite	铁质石英岩
Fersman	费尔斯曼
fertilizer minerals	化肥原料矿产
FESEM	场发射扫描电子显微镜
FFT	火焰熔化测试
fibrolite	发夕线石,细硅线石
fibrous aggregate	纤维状集合体
fibrous fracture	纤维状断口
ficelle	淡灰褐色,浅灰褐色
field capacity	田间持水量
field practice	野外实习
field *P-T* curve	野外 *P*-*T* 曲线
Field-Emission Scanning Electron Microscopy (FESEM)	场发射扫描电子显微镜
FIGM	流体包裹体地质温度显微镜
Fiji	斐济
filemot	枯叶色,褐色
filmy aggregate	被膜状集合体
FIMA	流体包裹体显微温度分析
fine feldspathic quartzarenite	细粒长石石英净砂岩
fine grained	细粒的
fine grained texture	细粒结构
fine sandstone	细砂岩
fine-grained	细粒的
fining	细粒化作用
Finland	芬兰
fire clay deposit	耐火粘土矿床
firn snow	粒雪
first order transition	一级相变

firth	河口湾
fissile	易分裂的，可分裂的，叶状的
fission-track dating	裂变径迹计时
fission-track retention age	裂变径迹保留年龄
fissure	裂隙
fissure eruption	裂隙[式]喷发
fissure vein	裂隙脉
fixed components	固定组分
fixism	固定论
Flame and Fusion Testing (FFT)	火焰熔化测试
flame structure	火焰状构造
flare spot	耀斑
flaser bedding	脉状层理，压扁层理，条纹层理
flattening	压扁[作用]
F-lattice	面心格子(F-格子)
flaxen	亚麻[黄]色，淡黄色
flesh-colored	肉色
flexibility	挠性
flexural-flow fold	弯流褶皱
flexural-slip fold	弯滑褶皱
flexure	挠曲
flint	燧石
flinty	燧石状
flocculation	聚沉作用，絮凝，凝聚
flood	洪水
flood period	汛期
flood plain	泛滥平原
Florida	佛罗里达州(美)
flow structure	流动构造
flower blue	花青
fluid convection	流体对流
fluid dynamics	流体动力学
fluid inclusion	流体包裹体

Fluid Inclusion Geothermometric Microscopy (FIGM)	流体包裹体地质温度显微镜
Fluid Inclusion Microthermometric Analysis (FIMA)	流体包裹体显微温度分析
fluid overpressure	流体超压
fluid pressure	流体压力
fluid viscosity	流体粘度
fluidal structure	流纹构造,流动构造
fluid-rock interaction	流体-岩石交互作用
fluorapatite	氟磷灰石
fluorescence	荧光
Fluorescence Microscopy (FM)	荧光显微镜
fluoride poisoning	氟中毒
Fluorine(F)	氟
fluorite	萤石
fluorite deposit	萤石矿床
fluorite type U-ore	萤石型铀矿
fluorosis	氟中毒
flute	槽,冲槽,凹槽
flute marks	槽痕
fluvial	河流的
fluvial action	流水作用
fluvial landform	流水地貌
fluxion structure	流状构造
flysch	复理石
FM	荧光显微镜
focal depth	震源深度
foehn wind	焚风
fog	雾
fold	褶皱
fold analysis	褶皱分析
fold generation	褶皱世代
fold geomorphology	褶皱构造地貌

fold interference pattern	褶皱干扰型式
folded mountain	褶皱山
folding	褶皱[作用]
foliated	叶片状
foliation	面理
food chain	食物链
food web	食物
footprint	足印,足迹
footwall	下盘
foraminifera	有孔虫类
force	力
forbidden band	禁带
fore-arc basin	弧前盆地
foreshock	前震
fore-shock	前震
form of environmental substance	环境物质形态
form of ore body	矿体形状
formation	组
formation age	形成年龄
formation volume factor	[原油]体积系数
forsterite	镁橄榄石
fossil fuels	化石燃料
fossiliferous	含有化石的
founding materials	铸造材料
Fourier synthesis	傅立叶合成
Fourier Transform Infrared Spectroscopy (FTIS)	傅立叶转换红外光谱
Fourier Transform Raman Spectroscopy (FTRS)	傅立叶转换拉曼光谱
fractional crystallization (FC)	分离结晶作用
fractionation curve	分馏曲线
fractionation effect	分馏效应
fractionation factor	分馏系数

fractionation mechanism	分馏机理
fracture	破裂,裂缝,断口,断裂
fracture cleavage	破劈理
fracture water	裂隙水
fracture zones	断裂带
fragment	碎屑
framestone	生物骨架灰岩
framework silicate	架状硅酸盐
France	法国
Francium(Fr)	钫
french blue	群青,暗青绿色,深蓝
french grey	浅灰色
French Guiana	法属圭亚那
French Polynesia	法属波利尼西亚
frequency distribution of element	元素分布频率
fringe	边缘,饰边,条纹
fringe alteration	边缘蚀变
front	锋面
frost	霜
frost splitting	冰劈作用
FTIS	傅立叶转换红外光谱
FTRS	傅立叶转换拉曼光谱
fugacity	逸度
fuller's earth	漂白土
fuller's earth deposit	漂白土矿床
fullerence	富勒烯(C_{60}晶体)
fullerite	富勒烯(C_{60}晶体)
fundamental intensity	基本烈度
fundamental particle	基本粒子
furongite	芙蓉铀矿
fusain	丝炭
fuscous	暗褐色
fused quartz deposit	熔炼石英矿床

fusinite	丝质体
fusinization	丝炭化作用
fusion crust	熔壳

G

gabbro	辉长岩
gabbro texture	辉长结构
Gabon	加篷
Gadolinium(Gd)	钆
Gaia hypothesis	盖雅假说
galactic plane	银道面
galactic year	银河年
Galaxy	银河[系]
galaxy	星系
galena	方铅矿
gallium deposit	镓矿床
Gallium(Ga)	镓
Gambia	冈比亚
Gamma-Ray Spectroscopy(GRS)	伽马射线光谱
gananite	赣南矿
gangue	脉石
gangue mineral	脉石矿物
garben texture	噶尔本结构(以柱状矿物定向分布于片理面上为特点)
garnet	石榴子石
garnet deposit	石榴石矿床
gas	气体
gas cap drive reservoir	气顶气驱储油层
gas coal	气煤
gas field	气田
gas from thermomaturation of organic matter	热成熟成因气
gas pool	气藏
gas retention age	气体保留年龄

gas saturation	含气饱和度
gas seepage	气苗
gash ore body	裂缝矿脉
gas-oil radio	气油比
gehlenite	钙铝黄长石
gelification	凝胶化作用
gem mineral	宝石矿物
gem mineralogy	宝石矿物学
gemology	宝石学
gemstone	宝石
general form	一般[单]形
generation	代,世代,发生
genetic mineralogy	成因矿物学
genetic type of mineral deposits	矿床成因类型
geobarometer	地质压力计
geobarometry	地质压力测定法
geobotanical method	地植物方法
geochemical anomaly	地球化学异常
geochemical background	地球化学背景
geochemical barrier	地球化学障
geochemical behaviour of element	元素地球化学行为
geochemical chart	地球化学图
geochemical classification of element	元素地球化学分类
geochemical closed system	地球化学封闭体系
geochemical cycle	地球化学旋回
geochemical cycle of uranium	铀地球化学旋回
geochemical detailed survey	地球化学详查
geochemical dispersion	地球化学分散
geochemical earth reference model(GERM)	地球化学参考模型
geochemical environment	地球化学环境
geochemical evolution	地球化学演化
geochemical facies	地球化学相
geochemical gas survey	气体地球化学测量

geochemical mapping	地球化学填图
geochemical mobility	地球化学活动性
geochemical model	地球化学模型
geochemical modeling	地球化学建模
geochemical open system	地球化学开放体系
geochemical prospecting	地球化学探矿
geochemical province	地球化学省
geochemical reconnaissance	地球化学普查
geochemical self-organization	地球化学自组织
geochemical survey of pollution	地球化学污染调查
geochemical system	地球化学体系
geochemical transport	地球化学输运
geochemical water survey	水地球化学测量
geochemistry	地球化学
geochemistry of element	元素地球化学
geochronologic scale	地质年代表
geochronologic unit	地质年代单位
Geochronology	地质年代学
geocryology	冻土地貌学
geode	晶腺
geodynamics	地球动力学
geography	地理学
geoid [surface]	大地水准面
geologic barometer	地质压力计
geologic body	地质体
geologic process	地质作用(过程)
geologic structure	地质构造
geologic time scale	地质年代表
geological boundary	地质边界
geological cycle	地质循环
geological event	地质事件
geological map	地质图
geological outcrop	地质露头

geology	地质[学]
geology of nonmetallic deposit	非金属矿床学
geology of ore deposits	矿床[地质]学
geomagnetic field	电磁场
geomagnetic pole	地磁极
geomagnetic reversal	地磁极性倒转
geomagnetism	地球磁性
geometric form	几何单形
geometrical crystallography	几何结晶学
geomorphic agency	地貌营力
geomorphic evolution	地貌演化
geomorphology	地貌学
Geophysical and Polar Research Center (GPRC)	地球物理与极地研究中心
geophysical technique	地球物理学方法
Georgia	佐治亚州(美),格鲁吉亚
geoscience	地学,地球科学
geosyncline	地槽
geosyncline-platform theory for metallogenesis	槽台成矿说
geologic thermometer	地质温度计
geotemperature	地温
geotherm	地热
geothermal geochemistry	地热地球化学
geothermal gradient	地热梯度
geothermal system	地热系统
geothermobarometer	地质温压计
geothermometer	地质温度计
geothermometry	地质温度测定法
GERM	地球化学参考模型
germanium deposit	锗矿床
Germanium(Ge)	锗
Germany	德国
geyserite	硅华
Ghana	加纳

giant planet	巨行星
Gibbs free energy	吉布斯自由能
gibbsite	三水铝石
ginger	淡赤黄色,姜黄色
glacial	冰川
glacial cirque	冰斗
glacial drift	冰碛
glacial erosion	冰蚀作用
glacial erratic boulder	冰川漂砾
glacial fluvial deposit	冰水堆积
glacial landform	冰川地貌
glacial pavement	冰溜面
glacial ploughing	刨蚀[作用]
glacial sapping	挖掘作用
glacial stage	冰期
glacial stria	冰川擦痕
glacial valley	冰蚀谷
glaciation	冰川作用
glacier	冰川
glacier lake	冰川湖
glaserite deposit	钾芒硝矿床
glass	玻璃,火山玻璃
glass raw materials	玻璃原料
glass shard	玻屑
glassy	玻璃质
glassy luster	玻璃光泽
glauconite	海绿石
glauconitic quartzarenite	海绿石石英砂岩
glaucophane	蓝闪石
glaucophane schist	蓝闪石片岩
glide	滑移[作用]
glide reflection plane	滑移面
glide twin	滑移双晶

global environment	全球环境
global warming	全球变暖
global sea level	全球海平面
glowing cloud	发光云,炽热火山云
gneiss	片麻岩
gneissose structure	片麻状构造
gneissosity	片麻理
gnomonic projection	心射极平投影
gobelin blue	暗青绿色
gobi	戈壁
geochemical exporation	地球化学勘查
geochemical rock survey	岩石地球化学测量
geochemical soil survey	土壤地球化学测量
goethite	针铁矿
gold	自然金
gold deposit	金矿床
gold placer	砂金矿床
Gold(Au)	金
golden [yellow]	金[黄]色
goldmanite	钙钒石榴子石
gold-quartz vein	含金石英脉
Goldschmidt V M	戈尔德施密特
Goldschmidt's rule	戈尔德施密特规则
Gondwanaland	冈瓦纳古大陆
goniometer	测角仪
goniometry	晶体测量
good cleavage	完全解理
gossan	铁帽
gossan-type iron deposit	铁帽型铁矿床
GPRC	地球物理与极地研究中心
graben	地堑
gradation	均夷[作用]
grade of ore	矿石品级

graded bedding	粒序层理,粒级层理，递变层理
graded sediment	级粒沉积,粒度递变沉积
graded surface	夷平面
Graduated Cylinder and Balance	分度圆筒与天平
grain	颗粒
grain flows	颗粒流
grain size	粒级
grained texture	颗粒结构
grainstone	粒状灰岩,粒屑灰岩，颗粒灰岩
grain-support	颗粒支撑
granite	花岗岩
granite type U-ore	花岗岩型铀矿
granitic gneiss	花岗质片麻岩
granitic migmatitic gneiss	花岗质混合片麻岩
granitic texture	花岗结构
granitization	花岗岩化[作用]
granitization theory of metallization	花岗岩化成矿说
granitoid	花岗岩类
granoblastic texture	花岗变晶结构
granoblastic-polygonal texture	花岗变晶多边形结构
granoblastic-polysutured texture	花岗变晶多缝合结构
granodiorite	花岗闪长岩
granophile deposit	亲花岗岩矿床
granophyre	花斑岩
granular	粒状
granule conglomerate	细砾岩
granulite	麻粒岩
granulite facies	麻粒岩相
graphic representation	图解表示
graphic solution	图解法
graphic texture	文象结构
graphite	石墨
graphite deposit	石墨矿床

grating texture	格子状结构
gravel	砾石，砾
Gravimetric Wet Chemical Analysis (GWCA)	测重湿化学分析法
gravitational process	重力[地质]作用
gravity	重力
gravity anomaly	重力异常
gravity cementation	重力式胶结
gravity drive reservoir	重力泄油储油层
gravity isostasy	重力均衡
gravity water	重力水
gray	灰色
grayish	浅灰色
grayish brown	灰棕色，灰褐色
grayish yellow	灰黄色
graywacke	杂砂岩
greasy luster	油脂光泽
Greece	希腊
green	绿色
green algae	绿藻
green blue	竹绿色
green schist	绿片岩
greenhouse effect	温室效应
greenhouse gas	温室气体
greenish	带绿色的
greenish blue	绿蓝色
Greenland	格陵兰(丹)
greenolith	铁蛇纹石
greenschist	绿片岩
greenschist facies	绿片岩相
greenstone	绿岩
greenstone belt	绿岩带
Greenwich time	格林威治时间
greisen	云英岩

greisenization	云英岩化
greisen-type tin deposit	云英岩型锡矿床
Grenada	格林纳达
grey	灰色
greywacke	杂砂岩,硬砂岩
grid twin	格子双晶
grind	磨[片]
grit	粗砂[岩]
groove	槽沟,细槽
gross primary production	总初级生产量
grossular	钙铝石榴子石
grossularite	钙铝石榴子石
ground fracture	地裂缝
ground mass	基质
ground moraine mound	底碛丘陵
ground moraine plain	底碛平原
ground setting	地面沉降
ground water	地下水,潜水
groundmass	基质
groundwater	地下水,潜水
groundwater-circulation metallogenic model	地下水循环成矿模式
group	族(矿物分类单位),群(地层单位)
growth	生长
growth hillocks	生长丘
growth steps	生长台阶
growth twin	生长双晶
GRS	伽马射线光谱
grunerite	铁闪石
Guadalupian	瓜德鲁普统
Guadeloupe	瓜德罗普岛(法)
Guam	关岛(美)
guano-type phosphate deposit	鸟粪石磷矿床
Guatemala	危地马拉

gugiaite	顾家石
Guinea	几内亚
Guinea-Bissau	几内亚比绍
Gulf steam	湾流
gupaiite	古北矿
Gutenberg discontinuity	古登堡不连续面
Guyana	圭亚那
GWCA	测重湿化学分析法
gymnosperm	裸子植物
gypsification	石膏化
gypsum	石膏
gypsum deposit	石膏矿床
gypsum plate	石膏试板

H

H	焓,热函
habitat	生活环境
hackly fracture	锯齿状断口
Hadean	冥古宙
hafnium deposit	铪矿床
Hafnium(Hf)	铪
hail print	雹痕
hailstone	冰雹
Haiti	海地
half-life	半衰期
halite	石盐,岩盐
halite deposit	石盐矿床
halle-flinta	细粒长英粒岩
halloysite	埃洛石
halogen family element	卤族元素
Hamersley-type iron deposit	哈默斯利型铁矿床
hand lens	放大镜
hand specimen	手标本

hanging valley	悬谷
hanging wall	上盘
Hard Mode Infrared Spectroscopy (HMIS)	硬模红外光谱
hardness	硬度
Hardness Testing (HDT)	硬度测试
harmotome	交沸石
harzburgite	斜方橄榄岩
hastingsite	绿钙闪石
hauyne	蓝方石
Hawaii	夏威夷州(美)
haydite materials	陶粒原料
hazard	自然灾害
hazel	淡褐色
HDAC	水热钻石高压盒
HDT	硬度测试
headward erosion	溯源侵蚀
health effect	健康效应
heat balance	热量平衡
heat capacity	热容
heat flow through the continent	大陆热流
heat flow through the ocean floor	海底热流
heat transfer process	热传导作用
heavy liquids	比重液
heavy metal pollution	重金属污染
heavy mineral	重矿物
heavy mineral deposit	重砂矿床
heavy oil	重质油
heavy rare earth element(HREE)	重稀土元素
hedenbergite	钙铁辉石
Heinlich event	海因里希事件
helicitic texture	残缕结构
heliogen blue	酞菁蓝
heliotrope	淡紫色

Helium(He)	氦
helvite	日光榴石
hematite	赤铁矿
hematitization	赤铁矿化
hemicrystalline	半晶质
hemidome	半坡面
hemihedron	半面体
hemimorphism	异极像
hemimorphite	异极矿
hemipelagic	半远洋的,近海的
hemipelagite	半远洋岩,近海岩
henna	红褐色,赤褐色,红褐色
Hercynian	海西期
hercynite	铁尖晶石
herringbone cross-bedding	羽状交错层理
heterodesmic lattice	多键型晶格
heterogeneous nucleation	非均匀成核
heterogranular texture	不等粒结构
heteropolar	异极
heterovalent isomorphism	异价类质同象
heulandite	片沸石
hexacelsian	钡霞石
hexagonal bipyramid	六方双锥
hexagonal close-packed	六方紧密堆积
hexagonal closest packing	六方最紧密堆积
hexagonal hemimorphic class	六方异极晶体
hexagonal prism	六方柱
hexagonal pyramid	六方单锥
hexagonal system	六方晶系
hexagonal trapezohedron	六方偏方面体
hexahedrite	方陨铁
hexahedron	六面体
hexakisoctahedron	六八面体

hexoctahedron	六八面体
HFSE	高场强元素
H-H_2O system	氢-水体系
hiatus	间断
hierarchical sampling	谱系采样
high aluminum mineral	高铝矿物
high field strength element(HFSE)	高场强元素(HFSE)
high m value(HIMU)	高 U/Pb 值的地幔(HIMU)
high pressure	高压
high relief	高突起
high white	高级白
high-alumina basalt	高铝玄武岩
high-grade gneiss	高级片麻岩
high-grade metamorphic rock	高级变质岩
high-Mg calcite	高镁方解石
High-Pressure Differential Thermal Analysis (HP-DAT)	高压差热分析
high-pressure facies series	高压相系
high-pressure metamorphic belt	高压变质带
high-pressure metamorphic rock	高压变质岩
High-Resolution Computed X-ray Tomography (HRCXT)	高分辨程序控制 X 射线层析
High-Resolution Electron Microscopy (HREM)	高分辨电子显微镜
High-Resolution Field-Emission Scanning Electron Microscopy (HR-FE-SEM)	高分辨场发射扫描电子显微镜
High-Resolution Neutron Powder Diffraction (HRNPD)	高分辨中子粉末衍射
High-Resolution Synchrotron X-ray Powder Diffraction (HRSXPD)	高分辨同步辐射 X 射线粉末衍射
High-Resolution Thermogravimetric Analysis (HRTGA)	高分辨热重分析

High-Resolution Transmission Electron Microscopy (HRTEM)	高分辨透射电子显微镜
High-Voltage Electron Microscopy (HVEM)	高电压电子显微镜
hill	丘陵
Hilt's rule	希尔特规律
Himalayan	喜马拉雅期
HIMU	高 U/Pb 值的地幔(HIMU)
hinge	枢纽
hinge line	枢纽线
hinge zone	枢纽带,转折端
HMIS	硬模红外光谱
H-O system	氢-氧体系
hoar	灰白色
Holmium(Ho)	钬
Holocene	全新统
Holocene Epoch(Series)	全新世(统)
holocrystalline	全晶质
holohedron	全面体
holohedral form	全形
holohedrism	全面像、全对称性
holotype	全型
homeostasis	内部自动平衡,动态平衡,自我平衡
homocline	单斜[层],同斜[层]
homodesmic lattice	单键型晶格
homogeneity	均一性
homogeneous	均匀的,均质的,均一的
homogeneous bedding	均匀层理
homogeneous deformation	均匀变形
homogeneous nucleation	均匀成核
homogenic	均质的
homogenic migmatite	均质混合岩
homogenization method	均一法
homogranular texture	等粒结构

homopolar	同极
homotype	同型
Honduras	洪都拉斯
honeycomb structure	蜂巢状构造,蜂窝状构造
hongshiite	红石矿
hopper crystal	漏斗晶
horizon	层位
horizontal bedding	水平层理
horizontal fold	水平褶皱
horizontal structure	水平构造
horizontal zoning	水平分带
horn	角峰
hornblende gneiss	角闪石片麻岩
hornblende	普通角闪石
hornblende hornfels	普通角闪石角岩
hornblende hornfels facies	普通角闪石角岩相
hornblende rock	角闪石岩
hornblende schist	角闪片岩
hornblendite	角闪石岩
hornfels	角岩
horse-stone	夹石
horsetail structure	马尾丝状构造
horst	地垒
host rock	容矿岩,主岩
hot brine reservoir	热卤水库
hot geothermal brine type deposit	地下热[卤]水型矿床
hot spot	热点
hot spring	热泉,温泉
hot spring deposit	热泉矿床
hot-brine metallogenesis	热卤水成矿作用
hourglass texture	砂钟构造
howardite	古铜钙长无球粒陨石
HP-DAT	高压差热分析

H-R diagram	赫-罗图(赫茨普龙-罗素图)
HREE	重稀土元素
HREM	高分辨电子显微镜
HR-FE-SEM	高分辨场发射扫描电子显微镜
HRNPD	高分辨中子粉末衍射
HRSXPD	高分辨同步辐射 X 射线粉末衍射
HRTEM	高分辨透射电子显微镜
HRTGA	高分辨热重分析
hsianghualite	香花石
huanghoite	黄河石
Hubble's law	哈勃定律
hue	主色调
huminite	腐植组(褐煤显微组分组)
humite	硅镁石
hummocky cross-bedding	丘状交错层理
humocollinite	无结构腐植体
humodetrinite	碎屑腐植体
humosapropelic coal	腐植腐泥煤(腐殖腐泥煤)
humotelinite	结构腐植体
humulite	腐植煤(腐殖煤)
humus	腐殖质
Hungary	匈牙利
Hungary blue	匈牙利蓝,钴蓝
hungchaoite	章氏硼镁石
hurricane	飓风
Huttenlocher intergrowth	休顿洛契连生(斜长石出溶形成)
HVEM	高电压电子显微镜
hyaloclastite	玄武碎屑岩,玻质碎屑岩,碎玻质熔岩
hyalomylonite	玻璃糜棱岩
hyalopilitic texture	玻晶交织结构
hybridrock	混染岩

hydration	水化(作用)
hydroastrophyllite	水星叶石
hydrocarbon	碳氢化合物,烃
hydrochlorborite	多水氯硼钙石
hydroclastic rock	水成碎屑岩
hydrodynamic force	水动力
hydrogen bond	氢键
hydrogen metasomatism	氢交代作用
Hydrogen(H)	氢
hydrogeochemistry	水文地球化学
hydrological cycle	水文循环
hydrolysed	水解的
hydrolysis	水解
hydromica	水云母
hydromicazation	水云母化
hydron blue	海昌蓝
hydroscopic water	吸附水
hydrosphere	水圈
hydrostatic pressure	静水压力
hydrothermal alteration	水热蚀变(热液蚀变)
hydrothermal channel	水热通道(热液通道)
hydrothermal deposit	水热矿床(热液矿床)
Hydrothermal Diamond-Anvil Cell (HDAC)	水热钻石高压盒
hydrothermal eruption	水热喷发
hydrothermal fluid	水热流体
hydrothermal reaction	水热反应
hydrothermal system	水热系统
hydrothermalism	热液作用
hydrous	水化的
hydrous melting curve	含水熔融曲线
hydroxide	氢氧化物
hydroxylapatite	羟磷灰石

hypabyssal intrusive facies	浅成侵入相
hypabyssal rock	浅成岩
hypautochthonous coal	微异地生成煤
hypergene mineral	表生矿物
hypersaline	超盐度
hypersthene	紫苏辉石
hypersthene granite	紫苏花岗岩
hypidioblastic	半自形变晶的
hypidiomorphic granular texture	半自形粒状结构
hypidiomorphic granular	半自形粒状
hypocenter	震源
hypogene anomaly	深成异常
hypothermal deposit	高温水热矿床(高温热液矿床)
hypothetical resource	远景资源

I

IAA	国际航天学会
IACMS	国际海洋科学咨询委员会
IACR	国际粘土研究协会
IAEG	国际工程地质学协会
IAG	国际大地测量学协会
IAGC	国际地球化学与宇宙化学协会
IAGOD	国际矿床成因协会
IAH	国际水文地质学家协会
IAHS	国际水文科学协会
IAMG	国际数学地质学协会
IAP	国际行星学协会
IAS	国际沉积学家协会
IAS	国际沉积学协会
IASH	国际科学水文学协会
IASPEI	国际地震学与地球内部物理学协会
IAU	国际天文学协会

IAV	国际火山学协会
IAVCEI	国际火山学与地球内部化学协会
ICCP	国际煤岩学委员会
ice age	冰期
ice blue	冰染蓝
ice cap	冰盖
ice raft	冰筏
Iceland	冰岛
ICES	国际海洋考察理事会
Iceland spar	冰洲石
ICG	联合会间地球动力学委员会
ichnofossils	痕迹化石
ichor	混合岩浆,岩汁
ICP	电感偶合等离子分析
ICP-MS	电感偶合等离子质谱仪
icsahedron	二十面体
ICSU	国际科学联合会理事会
icyl blue	衣色蓝
Idaho	爱达荷州(美)
idioblastic	自形变晶的
idiochromatic color	自色
idiomorphic granular texture	自形粒状结构
idocrase	符山石
IDTIMS	同位素稀释热离子化质谱仪
IFD	国际文献联合会
IFSEG	国际经济地质学家学会联合会
IGC	国际地质大会
igneous rock	火成岩
ignimbrite	熔结凝灰岩
IGS	国际冰川学会
IGU	国际地理学联合会
IISEE	国际地震学及地震工程学会
ijolite	霓霞岩

I-lattice	I-格子(体心格子)
Illinois	伊利诺伊州(美)
illite	伊利石
illite crystallinity	伊利石结晶度
ilmenite	钛铁矿
IMA	国际矿物学协会
imbricate fan	叠瓦扇
imbricate fault	叠瓦式断层
imbricate structure	叠瓦构造
immersion method	油浸法
immiscibility	不混溶性
immiscible magma	不混溶岩浆
immobile uranium	惰性铀
impact crater	陨[石撞]击坑
impact lava	冲击熔岩
impact marks	击象
impact metamorphic rock	冲击变质岩
impact metamorphism	冲击变质[作用]
impactite	冲击岩
impactite glass	冲击玻璃
imperfect cleavage	不完全解理
impure components	杂质组分
in situ	原地的
INAA	仪器中子活化分析
incarnadine	粉红色,肉色
incarnate	粉红色,肉色
incised meander	深切河曲
inclined bedding	倾斜层理
inclined fold	斜歪褶皱(倾斜褶皱)
inclosing rock	围岩
inclosure	包裹体
inclusion	包裹体
incommensurate structure	不相称结构

incompatible elements	不相容元素
incomplete isomorphous series	不完全类质同象
increasing zone of subsurface temperature	增温层
indanthrene blue	阴丹士林蓝
index mineral	指示矿物
index mineral zone	指示矿物带
index of crystal face	晶面指数
India	印度
Indiana	印第安纳州(美)
indicator element	指标元素
indicator plant	指示植物
indicatrix	光率体
indifferent components	孤立组分
indigo	靛[青]
indigo blue	靛蓝,紫蓝
indium deposit	铟矿床
Indium(In)	铟
indoine blue	吲哚因蓝
Indonesia	印度尼西亚
indophenol blue	靛酚蓝
Indosinian	印支期
induced earthquake	诱发地震
Inductively Coupled Plasma Analysis (ICP)	电感偶合等离子分析
Inductively Coupled Plasma-Mass Spectroscopy (ICP-MS)	电感偶合等离子质谱仪
industrial mineral	工业矿物
industrial minerals and rocks	工业矿物和岩石
inequigranular	不等粒状
inequigranular porphyritic texture	不等粒斑状结构
inert components	惰性组分
inert gas	惰性气体,亲气元素
inertinite	惰质组
inertite	微惰煤

inertodetrinite	碎屑惰质体
infiltration	渗滤
infiltration metasomatism	渗透交代[作用]
inflation	暴涨
Infrared Spectroscopy(IRS)	红外光谱
inherited argon	继承氩
inhomogeneous	非均质的
inhomogeneous deformation	非均匀变形
initial lead	初始铅
initial melt	初始熔融
initial ratio	初始比值
injected ore body	贯入矿体
injection gneiss	贯入片麻岩
injection migmatite	贯入混合岩
inland basin	内陆盆地
inner core	内核
inorganic genetic gas	无机成因气
inorganic geochemistry	无机地球化学
inorganic origin	无机起源
inorganic sedimentary structures	非生物沉积构造
inosilicate(chain silicate)	链状硅酸盐
impact crater	撞击坑
instantaneous isobars	瞬时等压线
instantaneous isotherms	瞬时等温线
Instrumental Neutron Activation Analysis (INAA)	仪器中子活化分析
insulator	绝缘体
integrity resources	资源的整体性
intensity	烈度
intercalation	插入,夹层
intercept of isochron	等时线截距
intercontinental rift	陆间裂谷
intercrystalline plastic deformation	晶界塑性变形

interfacial angle	面角
interference colors	干涉色
interference figure	干涉图
interference ripple	干涉波痕,干扰波痕
interglacial stage	间冰期
intergranular texture	拉斑[玄武]结构,间粒结构
intergrowth	交生,互生
interior basin	内陆盆地
interior chemistry of earth	地球内部化学
interlayer water	层间水
interlimb angle	翼间角
intermediate	中性
intermediate granulite	中性麻粒岩
intermediate rock	中性岩
intermediate-focus earthquake	中源地震
intermetallic compound	金属互化物
intermittent zoning	脉动分带,间歇分带
intermittent zoning	间歇分带
intermontane basin	山间盆地
internal bedding structure	层间构造
internal buffering	内缓冲
internal geomorphic agency	地貌内营力
internal isochron	内部等时线
internal reflection	内反射
internally buffered components	内缓冲组分
internally heated pressure vessel	内加热高压容器
International Academy of Astronautics (IAA)	国际航天学会
International Advisory Committee on Marine Sciences (IACMS)	国际海洋科学咨询委员会
International Association for Clay Research (IACR)	国际粘土研究协会

International Association for Mathematical Geology (IAMG)	国际数学地质学协会
International Association for Scientific Hydrology (IASH)	国际科学水文学协会
International Association of Engineering Geology (IAEG)	国际工程地质学协会
International Association of Geochemistry and Cosmochemistry (IAGC)	国际地球化学与宇宙化学协会
International Association of Geodesy (IAG)	国际大地测量学协会
International Association of Hydrogeologists (IAH)	国际水文地质学家协会
International Association of Hydrological Sciences (IAHS)	国际水文科学协会
International Association of Planetology (IAP)	国际行星学协会
International Association of Sedimentologists (IAS)	国际沉积学家协会
International Association of Sedimentology (IAS)	国际沉积学协会
International Association of Seismology and Physics of the Earth's Interior (IASPEI)	国际地震学与地球内部物理学协会
International Association of Volcanology (IAV)	国际火山学协会
International Association of Volcanology and Chemistry of the Earth's Interior (IAVCEI)	国际火山学与地球内部化学协会
International Association on Engineering Geology (IAEG)	国际工程地质学协会
International Association on the Genesis of Ore Deposits (IAGOD)	国际矿床成因协会
International Astronomical Union (IAU)	国际天文学协会
International Committee for Coal Petrology (ICCP)	国际煤岩学委员会

International Council for the Exploration of the Sea (ICES)	国际海洋考察理事会
International Council of Scientific Unions (ICSU)	国际科学联合会理事会
International Federation for Documentation (IFD)	国际文献联合会
International Federation of Societies of Economic Geologists (IFSEG)	国际经济地质学家学会联合会
International Geographical Union (IGU)	国际地理学联合会
International Geological Congress (IGC)	国际地质大会
International Glaciological Society (IGS)	国际冰川学会
International Institute of Seismology and Earthquake Engineering (IISEE)	国际地震学及地震工程学会
International Mineralogical Association (IMA)	国际矿物学协会
International Ocean Institute (IOI)	国际海洋学会
International Organization of Palaeobotany (IOP)	国际古植物学组织
International Paleontological Association (IPA)	国际古生物学协会
International Seismological Center (ISC)	国际地震学中心
International Society of Soil Science (ISSS)	国际土壤科学学会
International Subcommision on Stratigraphic Terminology (ISST)	国际地层术语分会
international symbol	国际符号(也称 HM 符号)
International Union for Quaternary Research (IUQR)	国际第四纪研究联合会
International Union of Biochemistry (IUB)	国际生物化学联合会
International Union of Biological Sciences (IUBS)	国际生物科学联合会
International Union of Crystallography (IUC)	国际结晶学联合会

International Union of Geodesy and Geophysics (IUGG)	国际大地测量与地球物理学联合会
International Union of Geological Sciences (IUGS)	国际地质科学联合会
International Union of Theoretical and Applied Mechanics (IUTAM)	国际理论力学与应用力学联合会
interparticle	粒间的
interplanar spacing	面网间距
interplanetary dust	行星际尘(埃)
interplanetary matter	行星际物质
intersecton lineation	交面线理
intersertal texture	填间结构
interstice	空隙
Interstitial	空隙的
intertidal zone	潮间带
Inter-Union Commission on Geodynamics (ICG)	联合会间地球动力学委员会
intraclast	内碎屑,内屑
intracratonic	克拉通内的,稳定陆块内的
intracrystalline plastic deformation	晶内塑性变形
intramicrite	内碎屑泥晶灰岩
intraplate strain	板内应变
intrasparite	内碎屑亮晶灰岩
intrusion	侵入[作用],侵入体
intrusive body	侵入体
intrusive contact	侵入接触
intrusive rock	侵入岩
invariant equilibrium	不变平衡
invariant point	不变点
invariant subgroup	不变子群
inverted metamorphic zones	倒转变质带
inversion	倒置(转)
iodine deposit	碘矿床

Iodine(I)	碘
IOI	国际海洋学会
ion exchange	离子交换
ionicity	离子性,电离度
ionisation potential	电离电势,电离电位
IOP	国际古植物学组织
Iowa	艾奥瓦州(美)
IPA	国际古生物学协会
Iran	伊朗
Iraq	伊拉克
Ireland	爱尔兰
iridescence	晕彩,晕色
iridium anomaly	铱异常
Iridium(Ir)	铱
iron deposit	铁矿床
iron meteorite	铁陨石
Iron(Fe)	铁
ironstone	铁岩,泥铁矿
IRS	红外光谱
ISC	国际地震学中心
island arc	岛弧
island arc rock	岛弧岩石
island texture	岛屿结构
island-arc system	岛弧系
isobar	等压线
isochemical	等化学的
isochemical metamorphic rock	等化学变质岩
isochemical metamorphism	等化学变质[作用]
isochron age	等时线年龄
isoclinal fold	等斜(同斜)褶皱
isodesmic structure	均键结构
isogon construction	等倾斜线作图法
isograd	等变线

isometric system	立方晶系,等轴晶系
isomorphic components	类质同象组分
isomorphic mixture	类质同象混合物
isomorphism	类质同象
isophysical metamorphic rock	等物理变质岩
isorank	等煤级线
isoreflectance line	等反射率线
isoseismal line	等震线
isostasy	均衡
isostasy theory	均衡说
isostatic anomaly	均衡异常
isostatic compensation	均量补偿,均衡补偿
isostructure	等结构,同结构
isotherm	等温线,等热
isotope dilution method	同位素稀释方法
Isotope Dilution Thermal Ionization Mass Spectroscopy (IDTIMS)	同位素稀释热离子化质谱仪
isotope fractionation factor	同位素分馏系数
isotope geochemistry	同位素地球化学
isotope geochronology	同位素地质年代学
isotope geothermometer	同位素地温计
isotopic equilibrium	同位素平衡
isotopic exchange equilibrium	同位素交换平衡
isotopic exchange reaction	同位素交换反应
isotopic fractionation	同位素分馏
isotopic geochronology	同位素地质年代学
isotropic	各向同性的
isotropic body	均质体
isotype	等型
isovalent isomorphism	等价类质同象
Israel	以色列
ISSS	国际土壤科学学会
ISST	国际地层术语分会

Italy	意大利
itabirite	铁英岩
itai-itai disease	痛痛病,骨痛病
Italian blue	意大利蓝,威尼斯蓝
IUB	国际生物化学联合会
IUBS	国际生物科学联合会
IUC	国际结晶学联合会
IUGG	国际大地测量与地球物理学联合会
IUGS	国际地质科学联合会
IUQR	国际第四纪研究联合会
IUTAM	国际理论力学与应用力学联合会
ivory	象牙[白]色,乳白色

J

jacinth	橘红色
jade	宝石
jade and nephrite deposit	玉石矿床
jade-green	浅绿色
jadeite	硬玉
Jamaica	牙买加
Japan	日本
jarosite	黄钾铁矾
jasmine	淡黄
JB	焦力天平(比重天平)
JCPDS card	粉末衍射标准卡片(JCPDS 卡)
jedrite	铝直闪石
jet	煤精,墨黑的
jet black	深黑,墨黑的
jetty	墨黑的
jinshajiangite	金沙江石
jixianite	蓟县矿
joint	节理

joint set	节理组
joint surface	节理面
joint system	节理系
Jolly Balance(JB)	焦力天平(比重天平)
Jordan	约旦
Jovian planet	类木行星
Jurassic	侏罗系
Jurassic Period(System)	侏罗纪(系)
juvenile mineral deposit	新生矿床

K

kali-feldspar gneiss	钾长片麻岩
kaliophilite	钾霞石
kalsilite	原钾霞石(六方钾霞石)
kamacite	铁纹石
Kambalda nickel deposit	卡姆巴尔达镍矿床
Kampuchea	柬埔寨
Kansas	堪萨斯州(美)
kaolin deposit	高岭土矿床
kaolinite	高岭石
kaolinization	高岭土化
K-Ar dating	钾-氩计时
Karahari manganese deposit	卡拉哈里锰矿床
karst	喀斯特,岩溶
karst cave	岩溶洞
karst lake	岩溶湖
karst landform	喀斯特地貌
karst process	喀斯特作用
karst topography	喀斯特地貌
karst water	岩溶水
karstfication	喀斯特作用
Kaschin-Beck disease	大骨节病
katazone	深成带

katoforite	红闪石
Kazakhstan	哈萨克斯坦
kelyphitic texture	次变边结构
Kempirsay chromium deposit	肯皮尔赛铬矿床
Kentucky	肯塔基州(美)
Kenya	肯尼亚
keratophyre	角斑岩
Kerch iron deposit	刻赤式铁矿床
kerogen	干酪根
kersantite	云斜煌岩
Keshan disease	克山病
key bed	标志层
key well	基准井,参数井
khaki	卡其色,黄褐色
khondalite	孔兹岩
kilchoanite	斜方硅钙石
kimberlite	金伯利岩
kimzeyite	钙锆石榴子石
kinematics	运动学
kink	扭折,膝折
kink band	扭折带
kink fold	膝折褶皱
kinking	扭折
Kirgizstan	吉尔吉斯斯坦
Kiribati	基里巴斯
Kiruna iron deposit	基鲁纳铁矿床
klippe	飞来峰
knee shaped twin	膝状双晶,肘状双晶
knife-edge crest	刃脊
komatiite	科马提岩
Krivoyrog-type iron deposit	克里沃罗格式铁矿床
Krypton(Kr)	氪
kupferschiefer type deposit	含铜页岩型矿床

kuroko deposit	黑矿型矿床
kuroshio	黑潮[日]
Kursk-type iron deposit	库尔斯克铁矿床
Kuwait	科威特
kyanite	蓝晶石
kyanite schist	蓝晶石片岩
kyanite-sillimanite type(or series)	蓝晶石-夕线石型(系列)

L

La Nina	拉尼娜现象
laboratory analysis	实验室分析
labradorescence	变彩
labradorite	拉长石
laccolith	岩盖
lacuna	缺失
lacustrine	湖泊的
ladder vein	梯状脉
lag grains	滞留颗粒
lagoon	泻湖,环礁湖
lahar	火山泥[石]流
laihunite	莱河矿
lake	湖泊
Lake Superior-type iron deposit	苏必利尔湖型铁矿床
LAM	激光磨蚀探针分析
lamella	叶片,纹理,条纹
lamina	细层,纹层,纹理
laminar flows	层流
lamination	分层, 纹理
lamprophyre	煌斑岩
lamprophyric texture	煌斑结构
land asphalt	地沥青
land desertization	土地沙漠化
Landau	兰多理论(郎道理论)

landscape geochemistry	景观地球化学
landform	地貌
landslide	滑坡
Lanthanum(La)	镧
Laos	老挝
lapilli	火山砾
lapilli tuff	[火山灰]凝灰岩
lapillistone	火山砾岩
large ionic lithophile element(LILE)	大离子亲石元素
larnite	斜硅钙石
Laser Isotope Analysis (LIA)	激光同位素分析
Laser Scanning Optical Microscopy (LSOM)	激光扫描光学显微镜
Laser Scattering (LS)	激光散射分析
Laser-Ablation Microprobe Analysis (LAM)	激光磨蚀探针分析
Laser-Ablation Sulfur Isotope Microprobe Spectroscopy (LASIM)	激光磨蚀硫同位素探针光谱
Laser-Ablation Sulfur Isotope Microscopy (LASIM)	激光磨蚀硫同位素显微镜
Laser-Fluorination Isotope Analysis (LFIA)	激光氟化同位素分析
Laser-Induced Fluorescence Spectroscopy (LIFS)	激光诱发荧光光谱
Laser-Interference Microscopy (LIM)	激光干涉显微镜
Laser-Microprobe Analysis (LMA)	激光探针分析
Laser-Raman Microprobe Spectroscopy (LRM)	激光拉曼探针光谱
Laser-Raman Microprobe Spectroscopy (LRMS)	激光-拉曼探针光谱
LASIM	激光磨蚀硫同位素探针光谱
LASIM	激光磨蚀硫同位素显微镜
late diagenesis	晚期成岩作用
late magmatic deposit	晚期岩浆矿床

late magmatic differentiation-type mineral deposit	岩浆晚期分异型矿床
late residual magma-type deposit	晚期残余岩浆型矿床
lateral blast	横向冲击波
lateral moraine bar	侧碛堤
lateral secretion theory	侧分泌说
laterite	砖红土
lateritization	红土化
laterized iron deposit	红土化型铁矿床
lath shaped	[长]板状
latitude effect	纬度效应
latosol	砖红土
lattice	晶格
lattice energy	晶格能
lattice point	晶格点
lattice preferred orientation	晶格优选方位
lattice site energy	晶格位能
Latvia	拉脱维亚
Laue group	劳厄群
laumontite	浊沸石
Laurasia	劳亚古大陆
laurel-green	浅[橄榄]绿色的
Laurentia	劳伦古大陆
lava	熔岩
lava dome	熔岩穹隆
lava flood	熔岩泛流
lava flow	熔岩流
lava fountain	熔岩喷泉
lava lake	熔岩湖
lava landform	熔岩地貌
lava plateau	熔岩高原
lava sheet	熔岩席,熔岩被
lava tube	熔岩管道

lava tunnel	熔岩隧道
lavender	淡紫色
lavender blue	淡紫兰色
law of Bravais	布拉维法则
law of constancy angle	面角守恒定律
law of crystal symmetry	晶体对称定律
law of rational indices	有理指数定律,整数定律
lawrencite	陨氯铁
Lawrencium(Lr)	铹
lawsonite	硬柱石
lawsonite-albite-chlorite facies	硬柱石-钠长石-绿泥石相
layer	岩层
layered structure	层状构造,带状构造
lazulite	天蓝石
lazurite	青金石
leaching	浸出,淋滤
leaching deposit	淋积矿床,淋滤矿床
leaching type uranium deposit	淋积型铀矿
lead white	铅白色
Lead(Pb)	铅
leadamalgam	汞铅矿
leaden	铅[灰]色
leadhat	铅帽
lead-zinc deposit	铅锌矿床
lead-zinc deposit in skarn	夕卡岩型铅锌矿床
leaf-like structure	叶片状构造
leaf-like texture	叶片状结构
leakage halo	渗漏晕
lean coal	瘦煤
lean gas	干气
Lebanon	黎巴嫩
LEED	低能电子衍射
left-handed crystal	左旋晶体

left-handed form	左形
legend	图例
Leitz-Jelley refractor	莱茨-杰利折射计
lemon	柠檬[黄]色
lemon-yellow	柠檬[黄]色
length fast, negative elongation	负延性
lenslike	透镜状
lensoid ore body	透镜状矿脉
lenticular bedding	透镜状层理
lepidoblastic texture	鳞片变晶结构
lepidolite	锂云母
leptite	长英麻粒岩
leptynite	变粒岩
Lesotho	莱索托
less good cleavage	不完全解理
leucite	白榴石
leucitite	白榴岩
leucocrate	浅色岩
leucocratic	淡色
leucocratic rock	浅色岩
leucoleptite	浅色长英麻粒岩
leucoleptynite	浅粒岩
leucosome	浅色体
LFIA	激光氟化同位素分析
lherzolite	二辉橄榄岩
LIA	激光同位素分析
Liberia	利比里亚
liberite	锂铍石
Libya	利比亚
lichen	地衣，苔癣
Liechtenstein	列支敦士登
LIFS	激光诱发荧光光谱
light	淡色

light blue	淡蓝
light bright yellow	浅亮黄
light brown	淡棕色,淡褐色
light green	淡绿色
light grey	浅灰色
light oil	轻质油
light orange	淡橙色
light rare earth element(LREE)	轻稀土元素
light-colored	淡色
light-colored mineral	浅色矿物
lightful	明亮的
lightness	色的亮度
lignite	褐煤
lilac	淡紫色,浅紫色
LILE	大离子亲石元素
linear flow structure	流线构造
LIM	激光干涉显微镜
limb	翼部
limestone	石灰石,石灰岩
limestone deposit	石灰岩矿床
limnic coal-bearing series	内陆型含煤岩系
limonite	褐铁矿
lineament	线状形迹,线性构造
linear fold	线状褶皱
linear productivity	线金属量
linear struscure	线性构造
lineation	线理
liptite	微壳煤(微稳定煤)
liptobiolite	残植煤(残殖煤)
liptodetrinite	碎屑壳质体(碎屑稳定体)
liquation deposit	熔离矿床
liquid immiscibility	液态不混溶作用
liquid window of oil generation	生油液态窗

liquidus	液相线
liquified natural gas(LNG)	液化天然气
listvenite	滑石菱镁[片]岩
litharenite	岩屑砂屑岩
lithic feldspar sandstone	岩屑长石砂岩
lithic fragment	岩屑
lithic pyroclast	[火山]岩屑
lithic quartz sandstone	岩屑石英砂岩
lithic sandstone	岩屑砂岩
lithic shards	岩屑
lithic wacke	岩屑杂砂岩
lithification	石化作用,岩化作用
lithium deposit	锂矿床
Lithium(Li)	锂
lithoclast	岩屑
lithofacies	岩相
lithologic trap	岩性圈闭
lithophile element	亲石元素
lithophysa structure	石泡构造
lithosized blue	石印蓝(无游离硫的群青,适用于油墨)
lithosome	岩体
lithosphere	岩石圈
lithosphere movement	岩石圈运动
lithosphere plate	岩石圈板块
lithostatic pressure	负荷压力
lithostratigraphic unit	岩石地层单位
Lithuania	立陶宛
litmus blue	石蕊蓝
lit-par-lit gneiss	间层片麻岩
lit-par-lit migmatite	间层混合岩
lizardite	利蛇纹石
Llandovery	兰多弗里统

liquefied flows	液化流
LMA	激光探针分析
LNG	液化天然气
load	负载
load casts	重荷模
loess liang	黄土梁
loess mao	黄土峁
loess plateau	黄土高原
loess yuan	黄土塬
local background	局部背景
local equilibrium	局部平衡
local metamorphism	局部变质[作用]
local time	地方时
lode	矿脉
loess	黄土
loess landform	黄土地貌
long-flame coal	长焰煤
longitude	经度
longitudinal dune	纵向沙丘
longitudinal wave(P wave)	纵波(P 波)
longitudinal zoning	纵向分带
lopezite	铬钾矿
Loping coal	乐平煤
Lopingian	乐平统
lopite	乐平煤
lopolith	岩盆
Lorraine-type iron deposit	洛林式铁矿床
Louisiana	路易斯安那州(美)
low angle reverse thrust	低角度逆冲断层
low relief	低突起
low temperature geochemistry	低温地球化学
low velocity zone	低速带
Low-Energy Electron Diffraction (LEED)	低能电子衍射

Lower hemisphere	下半球
Lower/Early Cambrian	下寒武统
Lower/Early Cretaceous	下白垩统
Lower/Early Devonian	下泥盆统
Lower/Early Jurassic	下侏罗统
Lower/Early Ordovician	下奥陶统
Lower/Early Triassic	下三叠统
low-grade metamorphic rock	低级变质岩
low-pressure facies series	低压相系
LRM	激光拉曼探针光谱
LRMS	激光-拉曼探针光谱
LS	激光散射分析
LSOM	激光扫描光学显微镜
Luanheite	滦河矿
Ludlow	拉德洛统
ludwigite	硼镁铁矿
luminescence	发光性
lunar core	月核
lunar crust	月壳
lunar geology	月球地质学
lunar lithosphere	月球岩石圈
lunar mantle	月幔
lunar meteorite	月球粒陨石
lunar regolith	月壤
lunar rock	月岩
lunar soil	月壤
lunar tectonics	月球构造学
lurain	月陆
luster	光泽
Lutetium(Lu)	镥
Luxembourg	卢森堡

M

MAC	活动元素吸收系数
Macedonia	马其顿
maceral	显微组分
macrinite	粗粒体(烟煤和褐煤显微组分)
Madagascar	马达加斯加
mafic	镁铁质
mafic mineral	铁镁矿物
mafic rock	镁铁[质]岩
maghemite	磁赤铁矿
magic number rule	幻数规则
Magic-Angle Spinning Nuclear Magnetic Resonance Spectroscopy (MAS-NMR)	变角自旋核磁共振光谱
magma	岩浆
magma chamber	岩浆房
magma conduit	岩浆通道
magma reservoir	岩浆房
magmatic differentiation	岩浆分异[作用]
magmatic injection-type deposit	岩浆贯入型矿床
magmatic mineral deposit	岩浆矿床
magmatic ore-forming process	岩浆成矿作用
magmatic-meteoric hydrothermal system	岩浆-大气水水热系统
magmatism	岩浆作用
magnesian	镁质
magnesian hornfels	镁质角岩
magnesian rock	镁质岩石
magnesioastrophyllite	镁星叶石
magnesiocummingtonite	镁闪石
magnesiohulsite	黑硼锡镁矿
magnesioriebeckite	镁钠闪石
magnesite	菱镁矿
magnesite deposit	菱镁矿矿床

magnesium deposit	镁矿床
Magnesium(Mg)	镁
magnetic anomaly	地磁异常
magnetic declination	磁偏角
Magnetic Force Microscopy (MFM)	磁力显微镜
magnetic inclination	磁倾角
magnetic storm	磁暴
magnetic susceptibility	磁化率
magnetism	磁性
magnetite	磁铁矿
magnetite quartzite	磁铁石英岩
magnitude of earthquake	地震震级
mahogany	红褐色,赤褐色,红褐色
main sequence	主星系(主序带)
main shock	主震
Maine	缅因州(美)
major element	主[量]元素
malachite	孔雀石
Malawi	马拉维
Malaysia	马来西亚
Maldives	马尔代夫
Mali	马里
malleability	延展性
Malta	马耳他
man-earth relationship	人-地关系
Manebach twin	曼尼巴双晶
manganese deposit	锰矿床
manganese deposit in weathering crust	风化壳型锰矿床
manganese hat	锰帽
manganese nodules on the ocean floor	海底锰结核
Manganese(Mn)	锰
manganese-hat-type manganese deposit	锰帽型锰矿床
manganite	水锰矿

Manitoba	马尼托巴省(加)
mantle	地幔
mantle convection	地幔对流
mantle geochemistry	地幔地球化学
mantle plume	地幔柱
mantle-derived	幔源[的]
manto	席状矿体(平卧矿体)
map interpretation	地质图解析
maple	淡棕色,淡褐色
marble	大理岩
marcasite	白铁矿
mare	月海
margarite	珍珠云母
marginal-sea basin	边缘海盆地
marialite	钠柱石
marine cave	海蚀洞(穴)
marine environment	海洋环境
marine geochemical exploration	海洋化探
marine geochemistry	海洋地球化学
marine sedimentary bauxite deposit	海相沉积铝土矿矿床
marine sedimentary manganese deposit	海相沉积锰矿床
marine sedimentary phosphorite deposit	海相沉积磷块岩矿床
marine-cut bench	海蚀阶地
marine-cut platform	波切台
marl	泥灰岩
maroon	栗色,深棕色,褐红色
marrine-built platform	海积阶地
marrine-built terrace	海积阶地
marrine-cut bench	海蚀阶地
marsh	沼泽
Martinique	马提尼克岛(法)
Maryland	马里兰州(美)
maskelynite	熔长石

MAS-NMR	变角自旋核磁共振光谱
mass extinction	集群绝灭
mass movement	块体运动
Massachusetts	马萨诸塞州(美)
massive aggregate	块状集合体
massive bedding	块状层理
massive dolomitization	块状白云岩化,厚层白云岩化
massive structure	块状构造
Mass-Spectroscopy (MS)	质谱仪
Masuda-Coryell diagram	增田-科里尔图解
material flow	物质流
matrix	基质,填隙物质,充填物,矩阵
matrix-support	基质支撑
maturation	成熟
mature	成熟
maturity	成熟度
maturity of organic matter	有机质成熟度
Matuyama polarity chron	松山极性时
Mauritania	毛里塔尼亚
Mauritius	毛里求斯
mauve	紫红
maximum principal stress	最大主应力
maximum resource potential	最大资源潜力
mazarine	深蓝色
mazzite	针沸石
MBRS	微束拉曼光谱仪
MBS	穆斯保尔光谱
meager coal	贫煤
mean square of weighted deviates(MSWD)	等时线拟合参数
mean stress	平均应力
meandering stream	曲流河
mechanical property	力学性质
mechanical twin	机械双晶,滑移双晶

mechanical weathering	机械风化
median relief	中突起
medium arkose	中粒长石杂砂岩
medium grained	中粒的
medium grained texture	中粒结构
medium sandstone	中砂岩
medium-grade metamorphic rock	中级变质岩
medium-pressure facies series	中压相系
medium-range forecast	中期预报
megacryst	巨晶,大晶体
meionite	钙柱石
meizoseismal area	极震区
mélange	混杂岩,混杂堆积
melanic	暗色
melanocrate	暗色岩
melanocratic	暗色
melanosome	暗色体
melilite	黄长石
meliphanite	蜜黄长石
melting	熔融
melting curve	熔融曲线
member	段
Mendelevium(Md)	钔
melanocratic rock	暗色岩
mercury deposit	汞矿床
Mercury(Hg)	汞
merwinite	默硅钙镁石
Mesoarchean	中太古界
mesocratic	中色的
mesocratic rock	中色岩
mesodesmic structure	中键结构
mesolite	中沸石
mesoperthite	中条纹长石

Mesoproterozoic	中元古界
Mesoproterozoic Era(Erathem)	中元古代(界)
mesosiderite	中铁陨石
mesosome	中色体
mesothermal deposit	中温水热矿床,中温热液矿床
Mesozoic Era(Erathem)	中生代(界)
Mesozoic(MZ)	中生界
mesozone	中带
meta-andesitization	变安山岩化,青盘岩化
metabasite	变基性岩
metabolism	新陈代谢[作用]
metacolloidal mineral	变胶体矿物
metacolloidal structure	变胶状构造
metacrystal ooid	变晶鲕
metacrystal texture	交代晶体结构
meta-gabbro	变辉长岩
metagalaxy	总星系
metal nickel-iron	金属镍铁
metallic luster	金属光泽
metallic mineral	金属矿物
metallic mineral deposit	金属矿床
metallic ore	金属矿产
metallization by metamorphic secretion	变质分泌成矿作用
metallogenesis	成矿作用
metallogenetic epoch	成矿时代
metallogenic belt	成矿带
metallogenic domain	成矿域
metallogenic epoch	成矿期
metallogenic epoch	成矿时代
metallogenic factor	成矿控制因素
metallogenic formation	成矿建造
metallogenic formation of diwa (geodepression) type	地洼型成矿建造

metallogenic formation of geosyncline type	地槽型成矿建造
metallogenic formation of platform type	地台型成矿建造
metallogenic megaprovince	成矿域
metallogenic model	成矿模式
metallogenic prediction	成矿预测
metallogenic province	成矿省
metallogenic region	成矿区
metallogenic series	成矿系列(成矿序列)
metallogenic specialization	成矿专属性
metallogeny	成矿学,矿床[成因]学,矿床成因说
metallogeny of deep lineaments	深部构造线成矿说
metallo-organic complex	金属有机络合物
metallo-organic matter mineralization	金属有机矿化作用
metamictization	蜕晶作用,变生非晶化,放射性损坏结构
metamorphic belt	变质带
metamorphic crystallization	变质结晶[作用]
metamorphic cycle	变质旋回
metamorphic differentiation	变质分异
metamorphic episode	变质阶段(幕)
metamorphic epoch	变质期
metamorphic event	变质事件
metamorphic facies	变质相
metamorphic facies group	变质相组
metamorphic facies series	变质相系
metamorphic field gradient	野外变质梯度
metamorphic grade	变质级
metamorphic metallization	变质成矿作用
metamorphic mineral deposit	变质矿床
metamorphic reaction	变质反应
metamorphic regime	变质体制
metamorphic rock	变质岩

metamorphic structure	变质构造
metamorphic subfacies	变质亚相
metamorphic texture	变质结构
metamorphic thermal gradient	变质热梯度
metamorphic zone	变质带
metamorphism	变质作用
metamorphogenic deposit	变成矿床(变质生成矿床)
metamorphosed cherty iron formation	受变质硅铁建造
metamorphosed deposit	受变质矿床
metamorphosed sedimentary iron deposit	受变质沉积铁矿床
metasomatic lead-zinc deposit in carbonate rock	碳酸盐岩铅锌交代矿床
metasomatic metamorphism	交代变质[作用]
metasomatic mineral deposit	交代矿床
metasomatic relict texture	交代残余结构
metasomatic rock	交代岩
metasomatism	交代[作用]
metasomatosis	交代[作用]
metastable	准稳的,亚稳的
metastable state	准稳[定状]态
metatexite	半熔岩
meteor	流星
meteoric waters	大气降水,雨水
meteorite	陨石
meteorite crater	陨石坑
meteorite full	陨石雨
meteorite impact	陨[石撞]击作用
meteorite-impact feature	陨击构造
meteoritics	陨石学
metric	公制的,米制的
Mexico	墨西哥
MFM	磁力显微镜
MHS	磨氏硬度计

miarolitic structure	晶洞构造
mica	云母
mica deposit	云母矿床
mica plate	云母试板
mica schist	云母片岩
Michigan	密歇根州(美)
micrinite	微粒体
micrite	泥晶灰岩,微晶灰岩(曾用名)
Micro Proton-Induced X-ray Emission Spectroscopy (m-PIXE)	显微光子诱导 X 射线发射光谱
Microbeam Raman Spectrometry (MBRS)	微束拉曼光谱仪
microbio-geological process	微生物地质过程
micro-biological cycle	生物小循环
microcline	微斜长石
microclinization	微斜长石化
microelement	微量元素
micrograined	微粒的
micrographic	显微文象结构的,显微照相的
Microindentation Hardness Testing (MIHT)	显微压痕硬度测试
microlite	细晶石,微晶
microlithon	微劈石
microlithotype	显微煤岩类型
microorganism	微生物
microperthite	显微条纹长石(光学显微镜下)
microphotometer	显微光度计
micropoikilitic	微嵌晶状的
Microprobe Analysis(MPA)	显微探针分析
Micro-Raman Spectroscopy(MRS)	显微拉曼光谱
microstructure	显微构造
microtektite	微玻璃陨石
Microthermometric Analysis (MTA)	显微热分析
mid night blue	深蓝色

Middle Cambrian	中寒武统
Middle Devonian	中泥盆统
Middle Jurassic	中侏罗统
Middle Ordovician	中奥陶统
Middle Triassic	中三叠统
mid-ocean ridge	[大]洋中脊
mid-ocean ridge rock	洋中脊岩石
mid-oceanic ridge	[大]洋中脊
mid-oceanic-ridge-basalt(MORB)	大洋中脊玄武岩
migmatic complex	混合杂岩
migmatism	混合岩化[作用]
Migmatite	混合岩
migmatitic gneiss	混合片麻岩
migmatitic granite	混合花岗岩
migmatization	混合岩化[作用]
migmatization metamorphic rock	混合岩化变质岩
migration of element	元素迁移
MIHT	显微压痕硬度测试
Milankovich theory	米兰科维奇学说
milk gold	乳金
milky opalescence	乳蛋白晕彩
milky quartz	乳石英
Miller's indices	米勒指数（米氏符号）
Miller-Urey reaction	米勒-尤里反应
Minamata disease	水俣病
Minas Gerais-type iron deposit	米纳斯吉拉斯型铁矿床
mine	矿山
mineragenetic series of ore deposits	矿床系列
mineral	矿物
mineral assemblage	矿物组合
mineral association	矿物组合
mineral chemistry	矿物化学
mineral deposit	矿床

mineral deposit by biogenic and biochemical sedimentation	生物化学沉积矿床
mineral deposit by cavity filling	充填矿床
mineral deposit by chemical sedimentation	化学沉积矿床
mineral deposit by colloidal agglutination	胶体化学沉积矿床
mineral deposit by mechanical sedimentation	机械沉积矿床
mineral deposit by weathering	风化矿床
mineral deposit geochemistry	矿床地球化学
mineral deposit geology	矿床地质学
mineral deposit model	矿床模式
mineral deposit	矿床
mineral deposits	矿床学
mineral economics	矿产经济学
mineral exploitation	矿产开发
mineral exploration	矿产勘查
mineral formula	矿物化学式
mineral fuel	燃料矿产
mineral fuel deposit	燃料矿床
mineral green	石绿
mineral isochron	矿物等时线
mineral lineation	矿物线理
mineral material for artware	工艺美术原料矿产
mineral material for chemical industry	化工原料矿产
mineral material for insulation	绝缘材料矿产
mineral paragenesis	矿物共生[组合]
mineral physics	矿物物理
mineral prospecting	矿产普查
mineral resource	矿产
mineral synthesis	矿物合成
mineral wool	矿棉
mineral yellow	石黄
mineralization	矿化作用
mineralization of U-Fe type	铀-铁型矿化

mineralization of U-Hg type	铀-汞型矿化
mineralization of U-P type	铀-磷型矿化
mineralization of U-Ti type	铀-钛型矿化
mineralizer	矿化剂
mineralizing fluid	矿化流体
mineralogical maturity	矿物成熟度
Mineralogical Mode Analysis (MMA)	矿物模式分析
mineralogical phase rule	矿物相律
mineralogy	矿物学
mineraloid	准矿物
minerogenesis	成矿作用
minerogenetic series	成矿系列(成矿序列)
minette	云煌岩
minimum internal energy	最小内能性
minimum principal stress	最小主应力
mining geology	矿山地质学
minium	朱红
Minnesota	明尼苏达州(美)
minnesotaite	铁滑石
minor element	次元素
Miocene	中新统
Miocene Epoch(Series)	中新世(统)
mirabilite	芒硝
mirabilite deposit	芒硝矿床
mire	沼泽
Mississippi	密西西比州(美)
Mississippian	密西西比亚系
Mississippi-valley[-type] lead-zinc deposit	密西西比河谷式铅锌矿床
Missouri	密苏里州(美)
misty gray	雾灰色
mixed crystal	混合晶体
mixed water	混合水
mixing model	混合模型

mizzonite	中柱石
MMA	矿物模式分析
mobile belt	活动带
mobile components	活动组分
mobile element absorption coefficient (MAC)	活动元素吸收系数
mobile uranium	活性铀
mobilism	活动论
mobility of element	元素活动性
mode	实际矿物成分
mode of occurrence of element	元素存在形式
mode of occurrence of ore body, attitude of ore body	矿体产状
model age	模式年龄
modena	深紫色
modern tectonic movement	现代构造运动
modulated structure	调制结构
module	模,阶
Moho	莫霍洛维奇不连续面(莫霍面)
Mohorovicic discontinuity (Moho)	莫霍洛维奇不连续面(莫霍面)
Mohs' hardness	摩氏硬度(摩斯硬度)
Mohs' hardness scale (MHS)	磨氏硬度计
Mohs' scale of hardness	摩斯硬度计
molar tooth (MT) structures	臼齿构造
Molar Volume Calculation (MV)	摩尔体积计算
molasse	磨拉石
mold pore	铸模孔
Moldova	摩尔多瓦
molds	铸模
molecular orbital theory	分子轨道理论
molecular organic geochemistry	分子有机地球化学
molecular sieve	分子筛
mollusk	软体动物

molybdenate	钼酸盐
molybdenite	辉钼矿
molybdenite-quartz vein	辉钼矿-石英脉
molybdenum deposit	钼矿床
Molybdenum(Mo)	钼
Monaco	摩纳哥
monalbite	单斜钠长石,蒙钠长石
monazite	独居石
Mongolia	蒙古
monocline	单斜[层]
monoclinic system	单斜晶系
monocrystalline	单晶
monogenic claystone	单成分粘土岩
monohydrate	单水型的,一水化合物
monomer	单分子物体,单聚物
monomineralic rock	单矿岩
monotropic transformation	不可逆的(单向的)转变
monsoon	季风
Montana	蒙大拿州(美)
monticellite	钙镁橄榄石
montmorillonite	蒙脱石
montmorillonitization	蒙脱石化
Montserrat	蒙特塞拉特岛(英)
monzonite	二长岩
monzonitic granite	二长花岗岩
monzonitic texture	二长结构
moon	月球
moonquake	月震
moonstone	月光石
moraine	冰碛
moraine terrace	冰碛阶地
MORB	大洋中脊玄武岩
mordenite	丝光沸石

Morocco	摩洛哥
morphology	形态,形貌
morphology of minerals	矿物形态
morphology of ore body	矿体形态
morphology of the earth surface	地表形态学
mosaic cementation	镶嵌式胶结
mosaic texture	镶嵌结构
moss green	苔绿色
Mossbauer Spectroscopy (MBS)	穆斯保尔光谱
mother rock	母岩
motif	结构基元
motility of resources	资源的变动性
mottled extinction	斑点消光
Mount Isa polymetallic deposit	芒特艾萨多金属矿床
mountain	山
mountain glacier	山岳冰川
mountain range	山脉
mountain system	山系
mounting medium	载体介质
Mozambique	莫桑比克
MPA	显微探针分析
m-PIXE	显微光子诱导 X 射线发射光谱
MRS	显微拉曼光谱
MS	质谱仪
MSWD	等时线拟合参数
MTA	显微热分析
mud bank	泥滩
mud volcano	泥火山
mud crack	泥裂
muddy texture	泥状结构
mudflow	泥流,泥石流
mudlump	泥火山,泥丘
mudrock	泥[状]岩

mullite 莫来石，多铝红柱石
multiple twinning 聚片双晶
multiplicative halo 累乘晕
multistage system 多阶段体系
multisystem 复式系统
multi-utility of resources 资源的多用性
muscovite 白云母
mushroom 蘑菇石
Mutusala manganese deposit 莫托沙拉锰矿床
MV 摩尔体积计算
mylonite 糜棱岩
mylonitic texture 糜棱结构
myrmekitic texture 蠕虫[状]结构

N

NAA 中子活化分析
nacarat 桔红色
naked eye 肉眼
Namibia 纳米比亚
nannoplankton 微型浮游生物
nanpingite 南平石
nappe 推覆体
nappe structure 推覆构造
native chromium 自然铬
native sulfur deposit 自然硫矿床
natrolite 钠沸石
natron lake 碱湖(苏打湖)
nattierblue 淡青色
natural coke 天然焦
natural disaster 自然灾害
natural environment 自然环境
natural gas 天然气
natural gas geochemistry 天然气地球化学

natural gas liquid	天然气液
natural hazard reduction	减灾
natural release	自然释放
Natural Remnant Magnetism Measurements (NRMM)	自然残留磁性测量
natural resource	自然资源
natural type of ore	矿石自然类型
Nauru	瑙鲁
navy blue	藏青,海军蓝,深蓝
ND	中子衍射
nearshore	近滨,近滨的
Nebraska	内布拉斯加州(美)
nebula	星云
nebular hypothesis	星云说
nebulite	云染[状混合]岩
nebulitic gneiss	云染片麻岩
nebulitic structure	云染状(星云状、阴影状)构造
neck	岩颈
needle	针状
negative crystal	负晶,空晶
negative anomaly	负异常
negative character	负光性
negative form	负形
negative ooid	负鲕
negative relief	负突起
nematoblastic texture	纤状变晶结构
Neoarchean	新太古界
neoblast	新生变晶
neocrystallization	新成结晶
Neodymium(Nd)	钕
Neogene	新近纪
Neogene Period(System)	新第三纪(系)

neomorphic	新生变形的，新生形的，再生成形的
Neon(Ne)	氖
Neoproterozoic	新元古界
Neoproterozoic Ⅲ	新元古Ⅲ系
Neoproterozoic Era(Erathem)	新元古代(界)
neosome	新成体
neotectonic movement	新构造运动
neotype	补型
Nepal	尼泊尔
nepheline	霞石
nephelinization	霞石化
nepheline syenite	霞石正长岩
nephelinite	霞石岩
Neptunium(Np)	镎
neritic sedimentary deposit	浅海相沉积矿床
nesosilicate	岛状硅酸盐
nest-like ore body	鸡窝状矿体
net	面网
net pay thickness	[产层]有效厚度
net primary production	净初级生产量
net slip	总滑距
Netherlands	荷兰
net-transfer reaction	净转移反应
network deposit	网脉
network-former cation	成网阳离子
network-modifier cation	变网阳离子
Neuberg blue	纽伯格蓝
Neumann line	诺依曼线
Neustad twin	内乌斯塔德双晶
neutral blue	中性蓝
neutralization	中和作用，中和反应
Neutron Activation Analysis (NAA)	中子活化分析

Neutron Diffraction(ND)	中子衍射
Neutron Powder Diffraction (NPD)	中子粉末衍射
neutron star	中子星
Nevada	内华达州(美)
New Brunswick	新不伦瑞克省(加)
New Caledonia	新喀里多尼亚(法)
New Hampshire	新罕布什尔州(美)
New Jersey	新泽西州(美)
New Mexico	新墨西哥州(美)
New York	纽约州(美)
New Zealand	新西兰
Newfoundland	纽芬兰省(加)
NHRL	北半球回归线
NHRL(nothern hemisphere regression line)	北半球回归线
Nicaragua	尼加拉瓜
niccolite	红砷镍矿
NICI	诺马尔斯基干涉相衬图
nickel deposit	镍矿床
Nickel(Ni)	镍
nickeliferous silicate deposit in weathering crust	风化壳型硅酸镍矿床
Niger	尼日尔
Nigeria	尼日利亚
Nikopol manganese deposit	尼科波尔锰矿床
niobium and tantalum deposit	铌钽矿床
niobium and tantalum deposit in altered granite	蚀变花岗岩铌钽矿床
Niobium(Nb)	铌
niobium-tantalum pegmatite deposit	铌钽伟晶岩矿床
niter	钾硝石
nitrate	硝酸盐
nitratine	钠硝石
nitratine deposit	钠硝石矿床(智利硝石矿床)

nitrification	硝化作用
nitrogen cycle	氮循环
Nitrogen(N)	氮
nitronatrite	钠硝石
Niue	纽埃(新)
NMRS	核磁共振光谱
Nobelium(No)	锘
noble gas	惰性气体
node	结点
nodular structure	瘤状构造
nodule	结核
Nomarski Interference Contrast Imaging (NICI)	诺马尔斯基干涉相衬图
non-conjugacy	不共轭
non-crystal	非晶体
non-crystallizing	非晶化
nondirectional structure	无定向构造
non-explosive eruption	非爆炸性喷发
nonferrous metal deposit	有色金属矿床
nonmetallic deposit	非金属矿床
non-metallic luster	非金属光泽
non-metallic mineral	非金属矿物
nonmetallic ore	非金属矿产
non-significant anomaly	无意义异常
non-skeletal	无骸晶的,无骨骼的,无骨架的
nonstoichiometry	非化学计量性
nontaxic nonepitaxial	非定向附生
nontronite	绿脱石
norite	苏长岩
normal epoch	正向期
normal fault	正断层
normal field	地磁正向场
normal stress	法向应力,正应力

normal succession	正常层序
normal twin	垂直双晶,正交双晶
normal zoning	顺向分带
normals to plane	法线
North Carolina	北卡罗来纳州(美)
North Dakota	北达科他州(美)
Northwest Territories	西北地区(加)
Norway	挪威
nosean	黝方石
Nova Scotia	新斯科舍省(加)
NPD	中子粉末衍射
NRMM	自然残留磁性测量
nuclear fuel waste	核燃料废物
nuclear geochemistry	核地球化学
Nuclear Magnetic Resonance Spectroscopy (NMRS)	核磁共振光谱
nuclear reaction	核反应
nuclear winter	核冬天
nucleation	成核
nucleic-acid-protein	核酸-蛋白质
nucleochronology	核年代学
nucleosynthesis in star	恒星原子核合成
nucleosynthesis in universe	宇宙原子核合成
nuée ardente	炽热火山云
nut-brown	胡桃色,深棕色,褐色

O

OA	光轴
oblique extinction	斜消光
oblique-slip-fault	斜向滑动断层
obsidian	黑曜岩
occurrence of rock	岩石的产状
ocean	海洋

ocean current	海流,洋流
ocean floor metamorphism	洋底变质[作用]
ocean red tide	海洋赤潮
ocean ridge metamorphic rock	洋中脊变质岩
ocean-floor metamorphic rock	洋底变质岩
ocean-grey	淡银灰色
oceanic rise	洋隆
oceanic-island-basalt(OIB)	洋岛玄武岩
octahedrite	八面体铁陨石
octahedron	八面体
octet	八连线、八重晶
odd-even predominance of n-alkanes	正烷烃奇偶优势
OES	光学发射光谱
offshore oil gas field	海上油气田
offwhite	灰白色
Ohio	俄亥俄州(美)
OIB	洋岛玄武岩
oil accumulation	石油聚集
oil column height	油柱高度(油藏含油高度)
oil field	油田
oil gas structure	含油气构造
oil gas trap	含油气圈闭
oil occurrence	石油产状
oil pool	油藏
oil primary migration	石油初次运移
oil production rate	石油产量
oil sand	油砂
oil saturation	含油饱和度
oil secondary migration	石油二次运移
oil seepage	油苗
oil shale	油页岩
oil-bearing area	含油面积
oil-bearing formation	含油地层

oil-bearing structure	储油构造
oil-gas contact	油气界面
oil-producing region	产油区
oil-water contact	油水界面
oily luster	油脂光泽
Oklahoma	俄克拉何马州(美)
Oklo phenomenon	奥克洛现象
Oligocene Epoch(Series)	渐新世(统)
oligoclase	奥长石
oligomict conglomerate	单成分砾岩
olivaceous	橄榄[绿]色
olive	橄榄[绿]色
olive green	橄榄[绿]色
olive-drab	深绿褐色的
olivenite	橄榄铜矿
olivine	橄榄石
olivine-bronzite chondrite	橄榄古铜球粒陨石
olivine-hypersthene chondrite	橄榄紫苏球粒陨石
OM	光学显微镜
Oman	阿曼
omeiite	峨眉矿
omphacite	绿辉石
oncoid	似核形石
oncolite	核形石
one circle goniometer	单圈测角仪
onshore oil gas field	陆上油气田
Ontario	安大略省(加)
oolite	鱼卵状石灰石,鲕粒岩
oolitic	鲕状的,鲕粒岩的
oolitic aggregate	鲕状集合体
oolitic limestone	鲕粒灰岩
oolitic structure	鲕状构造
oomicrite	鲕粒泥晶灰岩

oosparite	鲕粒亮晶灰岩
ooze	软泥
opal	蛋白石
opal deposit	蛋白石矿床
opalescence	乳光,蛋白光
opalization	蛋白石化
opaque	不透明
open form	开形
open system	开放系统
ophiolite	蛇绿岩
ophiolite complex	蛇绿岩套
ophiolite suite	蛇绿岩套
ophiolite texture	辉绿结构
opthalmite	眼球状混合岩
optic axis	光轴
optic orientation	光性方位
optical anormaly	光性异常
Optical Emission Spectroscopy (OES)	光学发射光谱
Optical Micriscopy(OM)	光学显微镜
optical property	光学性质
optical quartz deposit	光学石英矿床
Optical Scanning Microscopy (OSM)	光学扫描显微镜
optical sign	光性符号
optic-axial angle	光轴角
optic-axial plane	光轴面
optic-axis(OA)	光轴
orange	橘色,橙色
orange brown	橙棕色
orange red	橙红
orbicular structure	球状构造
orchid	淡紫色
order	有序,阶
order of interference color	干涉色级序

order parameter	有序参数
order structure	有序结构
order-disorder transformation	有序-无序转变
ordinary chondrite	普通球粒陨石
ordinary pyroclastic rock	正常火山碎屑岩
ordinary ray	[寻]常光
Ordovician	奥陶纪(系)
Ordovician Period(System)	奥陶纪(系)
ore	矿石,矿产
ore column	矿柱
ore bed	矿层
ore body	矿体
ore chimney	矿筒
ore deposit	矿床,金属矿床
ore deposit of dispersed elements	分散元素矿床
ore deposit of penta-element formation(ore deposit of Ni-Co-Ag-Bi-U formation)	五元素建造矿床(镍-钴-银-铋-铀建造矿床)
ore deposit of platinum group	铂族元素矿床
ore deposit of radioactive elements	放射性元素矿床
ore deposit of rare elements	稀有元素矿床
ore deposit of rare-earth elements	稀土元素矿床
ore district	矿区
ore field	矿田
ore formation	矿石建造(含矿建造)
ore magma	矿浆
ore magma iron deposit	矿浆型铁矿床
ore microscope	矿相显微镜
ore mineral	矿石矿物
ore mineralogy	矿石矿物学
ore nest	矿巢
ore occurrence	矿点
ore petrology	矿石学
ore pillar	矿柱

ore pipe	矿筒
ore pocket	矿囊
ore shoot	富矿体
ore zone	矿带
ore-forming element	成矿元素
ore-forming level	成矿壳层
ore-forming process	成矿作用
ore-forming process by remobilization	活化转移成矿作用
Oregon	俄勒冈州(美)
organic geochemistry	有机地球化学
orientation angle	方位角
orientation survey	试点测量
oriented section	定向切片
origin of element	元素起源
origin of humankind	人类起源
origin of organic molecules	有机分子起源
original crust of earth	原始地壳
orogenic belt	造山带
orogenic metamorphism	造山变质[作用]
orogeny	造山作用(运动)
orographical rain	地形雨
orosirian	造山系
orpiment	雌黄
orthite	褐帘石
orthoamphibolite	正角闪岩
orthoclase	正长石
orthoclase gneiss	正长片麻岩
orthoclasization	正长石化
orthogneiss	正片麻岩
ortholamprophyllite	斜方闪叶石
orthomagmatic mineral deposit	正岩浆矿床
orthometabasite	正变基性岩
orthometamorphic rock	正变质岩

orthometamorphite	正变质岩
orthopyroxene	斜方辉石
orthopyroxenite	斜方辉石岩
orthorhombic system	斜方晶系（又称正交晶系）
orthorhombic(rhombic)bipyramid	斜方双锥
orthorhombic(rhombic)disphenoid	斜方四面体
orthorhombic(rhombic)prism	斜方柱
orthorhombic(rhombic)pyramid	斜方单锥
orthoschist	正片岩
orthosilicate	正硅酸盐
OSM	光学扫描显微镜
Osmium(Os)	锇
OSPE(octahedron site preference energy)	八面体择位能
ostracode	介形虫
osumilite	大隅石
outcrop	露头
outer core	外核
outwash fan	冰水扇
outwash plain	冰水沉积平原
overburden	盖层
overgrowth	浮生(外延生长)
oversaturated rock	过饱和岩
overthrust fault	逆掩断层
overturned fold	倒转褶皱
oxbow lake	牛轭湖
oxidation	氧化
oxidation-reduction potential	氧化还原电位
oxide	氧化物
oxidized zone	氧化带
oxidized zone of sulfide deposit	硫化物矿床氧化带
oxidizing environment	氧化环境
oxybiotite	氧黑云母
oxygen coefficient	含氧系数

oxygen fugacity	氧逸度
Oxygen(O)	氧
oxyhornblende	氧角闪石
oyster white	乳白色
ozocerite	地蜡
ozone	臭氧
ozone depletion	臭氧耗竭
ozonosphere	臭氧层

P

P wave	纵波，初波(P 波)
P/T ratio type of metamorphism	变质作用 *P/T* 比类型
P/T ratio types of regional metamorphism (or metamorphic regions)	区域变质(或变质区)*P/T* 比类型
packet aggregate	束状集合体
packstone	泥晶颗粒灰岩，泥粒灰岩，泥粒状灰岩
pahoehoe lava	绳状熔岩
paired metamorphic belts	双变质带
Pakistan	巴基斯坦
palaeocrust of weathering	古风化壳
palaeocurrent analysis	古流分析
palaeogeography	古地理
Palaeoproterozoic Era(Erathem)	古元古代(界)
palaeosol	古土壤
Palaeozoic Era(Erathem)	古生代(界)
palagonite	橙玄岩，橙玄玻璃
pale	浅色
pale blue	淡蓝
pale bluish green	海蓝色，浅蓝绿色
pale brown	淡棕色，淡褐色
pale green	淡绿色
pale grey	浅灰色

pale red	浅红
pale yellow	浅黄
Paleoarchean	古太古界
Paleocene	古新统
Paleocene Epoch(Series)	古新世(统)
Paleogene	古近系
Paleogene Period(System)	老第三纪(系)
paleogeography of coal-bearing series	含煤岩系古地理
paleogeotemperature	古地温
paleomagnetic belt	古地磁条带
paleomagnetic field	古地磁场
paleomagnetic pole	古地磁极
paleomagnetism	古地磁[学]
Paleoproterozoic	古元古界
paleosome	古成体
paleotectonic movement	古构造运动
Paleozoic(PZ)	古生界
Palestine	巴勒斯坦
palimpsest structure	变余构造
palimpsest texture	变余结构
palingenesis	岩浆再生[作用]
palinspastic reconstruction	复原重建
palladium(Pd)	钯
Pallasite	橄榄陨铁
palygorskite	坡缕石
Panama	巴拿马
pangea	泛大陆
panidiomorphic granular	全自形粒状
pansy	紫罗兰色，紫色
panthalassa	泛大洋
paprika	红椒色
Papua New Guinea	巴布亚新几内亚
paraamphibolite	副角闪岩

paracrystalline(poorly crystalline) 类结晶的,亚结晶的(结晶程度不好的)
paragenesis 共生
paragenesis analysis 共生分析
paragenetic assemblage 共生组合
paragenetic mineral 共生矿物
paragenetic ore 共生矿
paragneiss 副片麻岩
paragonite 钠云母
Paraguay 巴拉圭
paralic coal-bearing series 近海型含煤岩系
paralic sedimentary deposit 海陆交替相沉积矿床
parallel bedding 平行层理
Parallel Electron Energy-Loss Spectroscopy (PEELS) 平行电子能失光谱
parallel extinction 平行消光,直消光
parallel fold 平行褶皱
parallel grouping 平行连晶
parallel hexahedron 平行六面体
parallel intergrowth 平行连晶
parallel twin 平行双晶
paramagnetism 顺磁性
para-massive structure 准块状构造
parametabasite 副变基性岩
parametamorphic rock 副变质岩
parametamorphite 副变质岩
parameter 参数，参量
paramorphism 副像
paraschist 副片岩
parasitic cone 寄生火山锥
parasitic fold 寄生褶皱
para-unconformity 平行不整合
pargasite 韭闪石

partial equilibrium	部分平衡
partial extraction	偏提取
partial melting	部分熔融
partial melting of mantle	地幔部分熔融
Particle-Induced X-ray Emission Spectroscopy (PIXE)	粒子诱导X射线发射光谱
parting	裂开,裂理,夹矸
parting plane	裂开面
partition coefficient	分配系数
passive margins basins	被动大陆边缘盆地
past global change	过去全球变化
pathfinder element	探途元素
patronite lode deposit	绿硫钒脉状矿床
Patterson diagram	帕特森图
Pauling's rules	鲍林法则
pay streak	富金线
PCM	相差显微镜
PDB standard(Pee Dee belemnite standard)	PDB标准(碳同位素标准,美国南卡罗纳白垩系皮迪组的美洲似箭石)
pea green	淡绿色
peach	桃[红]色
peachblow	桃[红]色
peach-colored	桃[红]色
peachy	桃[红]色
peacock blue	孔雀蓝
peak metamorphic condition	顶峰变质条件
pearl blue	浅蓝灰色
pearly luster	珍珠光泽
peat	泥炭
peat formation	泥炭化作用
pebble	中砾,砾石
pebble conglomerate	中砾岩

pectolite	针钠钙石
pedion	单面,端面
pedosphere	土壤圈
PEELS	平行电子能失光谱,光电子能损光谱
pegmatite	伟晶岩
pegmatite mineral deposit	伟晶岩矿床
pegmatite uranium deposit	伟晶岩型铀矿
pegmatitic	伟晶的
pegmatitization	伟晶作用
Peking blue	北京蓝(头青)
pelagic	远洋的,远海的,浮游的,深海的
pelagite	海底锰结核
pelite	泥质岩
pelitic	泥质
pelitic hornfels	泥质角岩
pelitic rocks	泥质岩石
pellet	团粒,球粒,粪球粒
pelmicrite	球粒微晶灰岩
pelletoid	似团粒
pelsparite	球粒亮晶灰岩
pendant cementation	悬挂式胶结
penecontemporaneous	准同生作用
peneplain	准平原
peneplanation	准平原化作用
penetrate twin	贯穿双晶,穿插双晶
pengzhizhongite	彭志忠石
pennine	叶绿泥石
penninite	叶绿泥石
Pennsylvania	宾夕法尼亚州(美)
Pennsylvanian	宾夕法尼亚亚系
pentagonal dodecahedron	五角十二面体
pentagonal icositetrahedron	五角三八面体

pentagonal trisoctahedron	五角三八面体
pentagonal tristetrahedron	五角三四面体
pentlandite	镍黄铁矿
perch ground water	滞水
percussion marks	击象
perfect cleavage	极完全解理
perfectly mobile components	完全活动组分
periclase	方镁石
pericline	肖钠长石
pericline law	肖钠长石律双晶
peridotite	橄榄岩
Period	纪
periodic bond chain(PBC)theory	周期性键链理论
peristerescence	晕彩,晕色
peristerite	晕长石
perlite	珍珠岩
perlite deposit	珍珠岩矿床
permeability	渗透性,渗透率
permeable bed	透水层
permeation gneiss	渗透片麻岩
permeation migmatite	渗透混合岩
Permian	二叠系
Permian Period(System)	二叠纪(系)
perovskite-titanomagnetite deposit	钙钛矿-钛磁铁矿矿床
perovskite	钙钛矿
perpetual resource	恒定性资源
perthite	条纹长石
perthitic texture	条纹结构
Peru	秘鲁
PES	光电子光谱
PESM	光电子光谱仪
pestilence	瘟疫
petalite	透锂长石

petrochemistry	岩石化学
petrogenesis	岩石成因论
petrogenetic grid	成岩格子
petrography	岩相学
petroleum geochemistry	石油地球化学
petroleum geology	石油地质学
petroleum province	含油气省
petroleum region	含油气区
petroliferous basin	含油气盆地
PGE	铂系元素(铂族元素,PGE)
pH(acidity-alkalinity)	酸碱度
phacolith	岩鞍
phanerocrystalline aggregate	显晶集合体
Phanerozoic Eon(Eonothem)	显生宙(宇)
Phanerozoic(PH)	显生宇
phase components	相组分
Phase Contrast Microscopy (PCM)	相差显微镜
phase transformation of mineral	矿物相变
pH-*Eh* diagram	pH -*Eh* 图解(酸碱度与氧化还原电位图解)
phengite	多硅白云母
phenocryst	斑晶
phenomenology	唯象学,现象学
Philippines	菲律宾
phillipsite	钙十字沸石
physical zonality	自然地带性
phlogopite	金云母
phonolite	响岩
phonolitic texture	响岩结构
phosphate	磷酸盐
phosphate deposit	磷[酸盐]矿床
phosphorescence	磷光
phosphorite deposit	磷块岩矿床

phosphorous deposit	磷矿床
Phosphorus(P)	磷
Photo Electron Energy-Loss Spectroscopy (PEELS)	光电子能损光谱
Photo Electron Spectroscopy (PES)	光电子光谱
photochemical smog	光化学烟雾
Photoelectron Spectromicroscopy (PESM)	光电子光谱仪
Photoemission Spectroscopy (XPES)	光子发射光谱
photoluminescence	光致发光
phreatic water	潜水,地下水
phyllic zone	绢英带
phyllite	千枚岩
phyllitic structure	千枚状构造
phyllosilicate	层状硅酸盐
phyllonite	千糜岩
physical geochemistry	物理地球化学
physical geology	普通地质学
physical model	物理模拟
physical weathering	物理风化
physic-geographical environment	自然地理环境
physic-geographical system	自然地理系统
physiognomy	地貌
physiology	生理学
picotite	铬铁尖晶石
picrite	苦橄岩
picritic	苦橄质
picture	图像,照片
piedmont glacier	山麓冰川
piemontite	红帘石
piezoelectric axis	电轴
piezoelectricity	压电性
piezoquartz deposit	压电石英矿床
pigeonite	易变辉石

PIGME	质子诱导 γ 射线发射光谱
pillow lava	枕状熔岩
pillow structure	枕状构造
pilot survey	试生产测量
pilotaxitic texture	交织结构
pinacoid	平行双面(板面);轴面
pink	粉红色,肉色
pisolite	豆石
pisolitic aggregate	豆状集合体
pisolitic structure	豆状构造
pistachio	淡黄绿色
piston-cylinder apparatus	活塞-缸筒设备
Pitcairn Islands	皮特凯恩群岛(英)
pitch-black	墨黑的
pitchblende	沥青铀矿
pitchstone	松脂岩
pitchy luster	沥青光泽
PIXE	粒子诱导 X 射线发射光谱
PIXES	光子诱导 X 射线发射光谱
placer	砂矿床
placer mineral	重砂矿物
plagioclase	斜长石
plagioclasite	斜长岩
plagiogneiss	斜长片麻岩
plagiogranite	斜长花岗岩
plain	平原
plain of accumulation	堆积平原
plain of denudation	剥蚀平原
planar flow structure	流面构造
planation	夷平[作用]
planation surface	夷平面
plane-polarization	平面偏振
plane-polarized light(ppl)	单偏光

planetary	行星地质学
plasma	等离子体
plastic behavior	塑性行为
plastic deformation	塑性变形
plastic flow	塑性流动
plasticity	塑性
plate	板块
plate boundary strain rate	板块边界应变速率
plate collision	板块碰撞
plate cross-bedding	板状交错层理
plate margin	板块边缘
plate movement	板块运动
plate subducting	板块俯冲
plate tectonics	板块构造
plateau	高原
plateau age	坪年龄
plateau glacier	高原冰川
plate-tectonic model for metallogenesis	板块构造成矿模式
platform	地台
platinum	自然铂
platinum group element(PGE)	铂系元素(铂族元素,PGE)
Platinum(Pt)	铂
platy	[薄]板状
play of color	变彩
playa	干盐湖
Pleistocene	更新统
Pleistocene Epoch(Series)	更新世(统)
pleochroic formula	多色性公式
pleochroic scheme	多色性公式
pleochroism	多色性
pleochroism brown-yellow	褐-黄色多色性
plessite	合纹石
plicate structure	皱纹构造

plicated structure	皱纹构造
Pliocene	上新统
Pliocene Epoch(Series)	上新世(统)
plug	岩塞
plume	[地幔]柱,热流柱
plunge	倾伏[角]
plunging fold	倾伏褶皱
plurifacial metamorphic rock	多相变质岩
pluton	深成侵入体,侵入岩,深成岩
plutonic intrusive facies	深成侵入相
plutonic rock	深成岩
plutonite	深成岩
Plutonium(Pu)	钚
PM	原始地幔
pneumato-hydrothermal deposit	气成-热液矿床
pneumato-hydrothermal metamorphic rock	气液变质岩
podiform	豆荚状
poikilitic	嵌晶状的,嵌晶结构
poikilitic texture	嵌晶结构,包含结构
poikiloblastic texture	嵌状变晶结构
poikilolblast	变嵌晶
poikilotopic texture	嵌晶结构
point bar	边滩,点沙坝
point group	点群
Poland	波兰
polar axis	极轴
polar wandering	极移
polar wandering path	极游移轨迹
polariscopy(PS)	偏光显微镜
polarity-reversal	极性反转
polarization	极化,偏振
polarized light	偏光
polarizing microscope	偏光显微镜

polarizing prism	偏光棱镜(尼科尔棱镜)
pole distance angle	极距角
pole of rotation	旋转极
polish	抛光
polished section	光片
pollution	污染
pollution dose	污染剂量
pollution index	污染指数
pollution sources	污染源
pollution type	污染类型
Polonium(Po)	钋
population	种群
polycrase	复稀金矿
polycrystalline	多晶的,聚晶的
polycyclic metallogeny	多旋回成矿作用
polygenetic and compound deposit	复成矿床(多因复成矿床)
polygenic claystone	复成分粘土岩
polygonal mosaic	多边形拼图结构
polyhalite deposit	杂卤石矿床
polyhedron	多面体
polyhydrat	烧石膏
polymer	聚合物
polymerization	聚合作用
polymerization model	聚合模型
polymetallic deposit	多金属矿床
polymetallic nodule	多金属结核
polymetamorphic rock	复变质岩
polymetamorphism	多期变质[作用]
polymict conglomerate	复成分砾岩
polymineralic rock	复矿岩
polymorph	同质多象变体
polymorphic transformation	同质多象转变
polymorphic transition	多型转变

polymorphism	同质多象
polysaccharide	多糖
polysomatic	多晶体的,多晶畴的
polysome	多晶体现象,多晶畴现象
polysynthetic twin	聚片双晶
polytype	多型
poor cleavage	不完全解理
population explosion	人口爆炸
porcelain streak plate	瓷板
pore	孔隙
porosity	多孔性，有孔性,孔隙度
porosity water	孔隙水
porous cementation	孔隙式胶结
porphyrite	玢岩
porphyrite iron deposit	玢岩铁矿床
porphyritic	斑状
porphyritic iron deposit	玢岩铁矿床
porphyritic texture	斑状结构
porphyroblast	变斑晶
porphyroblastic texture	斑状变晶结构
porphyroclast	碎斑
porphyry	斑岩
porphyry copper deposit	斑岩铜矿床
porphyry copper model	斑岩铜矿模式
porphyry molybdenum deposit	斑岩钼矿床
Portugal	葡萄牙
positive character	正光性
positive form	正形
positive relief	正突起
post-glacial period	冰后期
postive anomaly	正异常
postmagmatic hydrothermal solution	岩浆期后热液
post-magmatic mineral deposit	岩浆期后矿床

Postmasburg manganese deposit	波斯特马斯堡锰矿床
post-tectonic blast	构造后变晶
potash feldspathization	钾长石化
potash salt deposit	钾盐矿床
potassic metasomatism	钾交代[作用]
potassic zone	钾化带
potassic-metasomatism type uranium deposit	钾交代型铀矿
Potassium(K)	钾
potential acidity	潜在酸度
powder blue	浅灰蓝色
powdery aggregate	粉末状集合体
ppb (part per billion)	十亿分之一(10^{-9})
ppl	单偏光
ppm(part per million)	百万分之一(10^{-6})
ppt(part per trillion)	万亿分之一(10^{-12})
primary inclusion	原生包体
Praseodymium(Pr)	镨
Precambrian	前寒武纪
precession	岁差
precious metal deposit	贵金属矿床
precious stone	宝石矿产
precipitation	沉淀[作用]
precipitation zoning	沉淀分带
precursor compound	生源物
prediction	预报
preferential nucleation	优先成核
preferred orientation	优选方位
prehnite	葡萄石
prehnite-pumpellyite facies	葡萄石-绿纤石相
preliminary prospecting well	预探井
PREMA	流行地幔
pressure calibration	压力校正

pressure fringe	压力裙
pressure gradient	压力梯度
pressure shadow	压力影
pressure solution	压溶
pressure transmitting medium	传压介质
pressure vessel	压力仓
pre-tectonic blast	构造前变晶
prevalent	普遍的,流行的
prevalent mantle(PREMA)	流行地幔
prevention	预防
Pridoli	普里多利统
primary anomaly	原生异常
primary environment	原生环境
primary halo	原生晕
primary joint	原生节理
primary mineral	原生矿物
primary production	初级生产
primary sedimentary structure	原生沉积构造
primary twin	原生双晶
primary wave(P wave)	初波(P 波)
primeval lead	原生铅
primeval soup	原始汤
primitive isotopic anomaly	原始同位素组成异常
primitive mantle(PM)	原始地幔
primordial lead	原始铅
primrose yellow	樱草黄,淡黄
Prince Edward Island	爱德华王子岛省(加)
principal axes of strain	应变主轴
principal optic axis	光学主轴
principal stress	主应力
principle	原理,法则
principle refractive index	主折射率
printing blue	印染蓝,印刷蓝

prism	柱
prism twin	柱面双晶
prismatic	柱状
prochlorite	蠕绿泥石
producer	生产者
producing well	生产[油]井
profile	正交剖面图,剖面
profile sampling	测线采样
prograde metamorphic rock	进变质岩
prograde metamorphism	进变质
progressive deformation	渐进变形,递进变形
progressive metallogenesis	递进成矿说
progressive metamorphism	递增变质
proluvial fan	洪积扇
proluvial placer	洪积砂矿
Promethium(Pm)	钷
protore	矿胎
propylite	青磐岩
propylitization	青磐岩化
prospect	远景区
prospecting mineralogy	找矿矿物学
Protactinium(Pa)	镤
protein	蛋白质
Proterozoic (PR)	元古宇
Proterozoic Eon(Eonothem)	元古宙(宇)
protist	原生生物
protocataclasite	初碎裂岩
protohydrosphere	原始水圈
protolith	原岩
protomylonite	初糜棱岩
Proton-Induced Gamma-ray Emission Spectroscopy (PIGME)	质子诱导 γ 射线发射光谱

Proton-Induced X-ray Emission Spectroscopy (PIXES)	质子诱导 X 射线发射光谱
proton luminescence	质子发光
protostar	原恒星
proven reserve	探明储量
provenance	起源,根源;陆源区,蚀源区
proximal ore deposit	近火山活动金属矿床
prussian blue	普鲁士蓝
PS	偏光显微镜
pseudochromatic color	假色
pseudoisochron	假等时线
pseudoleucite	假白榴石
pseudomorph	假晶,假象
pseudomorphic	假晶(假象)的
pseudomorphic of serpentine after olivine	橄榄石的蛇纹石假象
pseudomorphism	假象
pseudopeloid	假团粒
pseudo-secondary inclusion	假次生包体
pseudotachylite	假玄武玻璃
psilomelane	硬锰矿
P-T-t path	*P-T-t* 轨迹
ptygmatic migmatite	肠状混合岩
ptygmatite	肠状岩
puce	深褐色
Puerto Rico	波多黎各岛(美)
pulsation theory	脉动说
pulsative zoning	脉动分带,间歇分带
pumice	浮岩,浮石
pumpellyite	绿纤石
purple	紫色
purple black	紫黑色
purple brown	紫棕色,紫褐色
purplish	略带紫色的

purplish blue	藏蓝,紫蓝
purplish grey	紫灰色
purplish red	紫红
purply	略带紫色的
putrefaction	腐泥化作用
pycnometer	比重瓶
pyrolusite	软锰矿
pyramid	[单]锥
pyramidal dune	金字塔形沙丘
pyrite	黄铁矿
pyrite deposit	黄铁矿矿床
pyritiferous copper deposit in volcanic rocks	火山岩黄铁矿型铜矿床
pyritiferous polymetallic deposit in volcanic rocks	火山岩黄铁矿型多金属矿床
pyritization	黄铁矿化
pyritohedron	五角十二面体
pyroclast	火山碎屑
pyrochlore	烧绿石
pyrochlore-bearing carbonatite	烧绿石碳酸岩
pyroclastic	火山碎屑的
pyroclastic cone	火山碎屑锥
pyroclastic flow	火山碎屑流
pyroclastic rock	火山碎屑岩
pyroclastic texture	火山碎屑结构
pyroelectricity	热电性
pyrolite	地幔岩
pyrometamorphic rock	高热变质岩
pyrometamorphism	高热变质[作用]
pyrometasomatic deposit	高热交代矿床
pyrometasomatic rock	高热交代岩
pyrope	镁铝石榴子石
pyrophyllite	叶蜡石

pyrophyllite deposit	叶蜡石矿床
pyrophyllitization	叶蜡石化
pyroxene	辉石
pyroxene gneiss	辉石片麻岩
pyroxene hornfels	辉石角岩
pyroxene-hornfels facies	辉石角岩相
pyroxene-plagioclase stony-iron	中铁陨石
pyroxenite	辉石岩
pyroxenoid	似辉石
pyroxenolite	辉岩
pyroxferroite	三斜铁辉石
pyrrhotite	磁黄铁矿

Q

Qatar	卡塔尔
qingheiite	青河石
qitianlingite	骑田岭矿
quantitative model	定量模拟
quartz	石英
quartz andesite	石英安山岩
quartz arenite	石英净砂岩
quartz diorite	石英闪长岩
quartz exhalite	石英喷流岩
quartz monzonite	石英二长岩
quartz porphyry	石英斑岩
quartz sand deposit	石英砂矿床
quartz sandstone	石英砂岩
quartz trachyandesite	石英粗安岩
quartz trachyte	石英粗面岩
quartz wedge	石英楔
quartz-feldspathic	长英质
quartzification	石英化
quartzite	石英岩

quartzitic conglomerate	石英岩砾岩
quartz-keratophyre	石英角斑岩
quasicrystal	准晶体,准晶态
Quaternary	第四系
Quaternary ice age	第四纪冰期
Quaternary Period(System)	第四纪(系)
quaternarysystem	四元[体]系
Quebec	魁北克省(加)
quench	淬火
quench melt	淬火熔体

R

RAC	相对吸收系数
radial ooid	放射鲕
radiance	深粉红色
radiated	放射状
radiating aggregate	放射状集合体
radiation	辐射
radiation balance	辐射平衡
radioactive decay	放射性衰变
radioactive measurement	放射性测量
radioactive pollution	放射性污染
radiocarbon dating	放射性碳计时
radio-element	放射性元素
radiolarian	放射虫
radiolarite	放射虫岩
radiothermoluminescence	辐射热发光
radiotracer analysis	放射性示踪分析
Radium(Ra)	镭
radius	半径
radon measurement	氡法测量
Radon(Rn)	氡
rain	雨

rain mark	雨痕
raindrop print	雨痕
Raman Spectroscopy(RS)	拉曼光谱
ramification migmatite	分枝混合岩
random sampling	随机采样
rankinite	硅钙石
rapakivi texture	环斑结构
rare alkaline metal	稀有碱金属
rare earth element(REE)	稀土元素
rare earth metal deposit	稀土金属矿床
rare element	稀有元素
rare gas	惰性气体
rare-earth-bearing carbonatite deposit	稀土碳酸岩矿床
rare-earth-bearing magnetite-hematite deposit	稀土-磁铁矿-赤铁矿矿床
rational face	有理指数晶面
rattan yellow	藤黄,浅亮黄色
raw material for building and construction	建筑材料矿产
raw material for cement	水泥原料矿产
raw material for ferrous metallurgy	黑色冶金辅助原料矿床
raw material for glass	玻璃原料矿产
raw material for moulding	造型原料矿产
raw material for optical use	光学原料矿产
raw material for piezoelectricity	压电原料矿产
Rayleigh fractionation	瑞利分馏
Rb-Sr dating	铷-锶计时
reaction isograd	反应等变线
reaction rate	反应速率
reaction rim	反应边
reaction rim texture	反应边结构
reaction series	反应系列
reaction texture	反应结构
reactions that involve H_2O and/or CO_2	水和二氧化碳参加的反应

realgar	雄黄
realgar deposit	雄黄矿床
reciprocal lattice	倒易格子
reclined fold	斜卧褶皱
reconstructive transformation	重建型转变,重建式相变
recoverable resource	可采资源
recovery	恢复
recrystallization	再结晶,重结晶
recrystallize	再结晶,重结晶
recumbent fold	平卧褶皱
red	红色
red algae	红藻
red giant	红巨星
red shift	红移
red soil	红土
red-color alteration	红色蚀变
reddish	微[带]红[色]的
reddish blue	红蓝色,品蓝
reddish brown	红棕色
reddish orange	红橙色
redeposition	再沉积作用
redistribute	重新分配,再区分,重新分布
redox transitional zone of uranium	铀氧化-还愿过渡带
reduce	减少,缩小,简化,还原
reducing environment	还原环境
REE	稀土元素
reef	暗礁,生物礁,矿层
reef trap[structure]	礁圈闭[构造]
reflectance	反射,反射率
reflectance analysis	反射率分析法
reflectance imagine(RI)	反射像
reflectance refractometry(RR)	反射折射仪
reflection	反射

reflection color index	反射颜色指数
reflection dihedron	反映双面
reflectivity	反射率
reflectometer	反射计
refraction	折射
refractive index	折射率,折光率
refractory	难熔的,耐火材料
refractory materials	耐火材料
refringence	折光率,折射率
regenerated deposit	再生矿床
regenerated granite	再生花岗岩
regional background	区域背景
regional differentiation	地域分异
regional dynamic metamorphism	区域动力变质[作用]
regional dynamometamorphic rock	区域动力变质岩
regional dynamothermal metamorphism	区域动热变质[作用]
regional environment	区域环境
regional geochemical anomaly	区域地球化学异常
regional geochemical survey	区域地球化学测量
regional geology	区域地质学
regional metallogeny	区域成矿学
regional metamorphic rock	区域变质岩
regional metamorphism	区域变质[作用]
regional migmatization	区域混合岩化[作用]
regional static metamorphism	区域静力变质[作用]
regionalization of resources	资源的地域性
regolith	表皮土，风化层，表土
regression	海退
reinerite	砷锌矿
rejuvenated landform	地形侵蚀回春
rejuvenation of stream	河流回春
relative absorption coefficient(RAC)	相对吸收系数
relative abundance	相对丰度

relative density	相对密度
relative permeability	相对渗透率
release of environmental substance	环境物质释放
relict structure	残余构造
relict texture	残余结构
relief	突起
remnant magnetism measurements (RMM)	残留磁性测量
remobilization	重新活动,再活化作用,活化转移
reniform aggregate	肾状集合体
reniform structure	肾状构造
replacement	交代[作用]
reptilia	爬行类
reptilian age	爬行动物时代
Republic of Korea	韩国
Republic of the Marshall Islands	马绍尔群岛共和国
reseda	浅绿色
reserve	储量
reservoir	储集层,油层,产油层,水库
reservoir formation	储油气层
reservoir geology	储油地质学
reservoir property	储层性质
residence time	居留时间
residual and mechanical concentration	残积和机械富集作用
residual deposit	残积矿床,残留矿床(风化壳矿床)
residual sediments	残积物
resinite	树脂体
resinous luster	松脂光泽,树脂光泽
resonant ionization mass-spectrometry (RIMS)	共振离子化质谱仪
resource base	资源基础
resource scarcity	资源短缺

resource scarcity due to environmental degradation	环境退化造成的资源短缺
resource scarcity due to geopolitics	地缘政治造成的资源短缺
resource scarcity due to under-development	欠发达造成的资源短缺
resource shortage	资源短缺
resources	资源
resources assessment	资源评价
reticular density	面网密度
retrograde metamorphic rock	退变质岩
retrograde metamorphism	退变质
retrogressive metamorphism	退化变质
Reunion	留尼汪岛(法)
reversal of geomagnetic field	电磁场倒转
reverse fault	逆断层
reverse magnetization	反磁化
reverse zoning	反向环带,逆向分带
reversed epoch	反向期
reversed field	地磁负向场
reversed succesion	倒转层序
revolution	公转
rhenium deposit	铼矿床
Rhenium(Re)	铼
rheology	流变学,河流学
Rhode Island	罗得岛州(美)
Rhodium(Rh)	铑
rhodochrosite	菱锰矿
rhodonite	蔷薇辉石
rhodonite deposit	蔷薇辉石矿床
rhombic dodecahedron	菱形十二面体
rhombododecahedron	菱形十二面体
rhombohedral system	菱面体晶系,三方晶系
rhombohedron	菱面体
Rhyacian	层侵系

rhyolite	流纹岩
rhyolitic	流纹质
rhyotaxitic texture	流纹结构
RI	反射像
ribben texture	丝带结构
rich gas	富气
richterite	钠透闪石
riebeckite	钠闪石
Rietveld X-ray Modal Analysis Method (RXMAM)	若特维尔德 X 射线矿物含量分析方法(?)
rift	裂谷,裂陷
rift geomorphology	断裂地貌
right-handed crystal	右旋晶体
right-handed form	右形
right way up	正常层序
RIMS	共振离子化质谱仪
ring silicate	环状硅酸盐
ring-shaped twin	环状双晶
Ringwood A E	林伍德
Ringwood's rule	林伍德规则
ripple	波痕,浅滩
ripple crest	波峰
ripple mark	波痕
risk assessment	风险评价
river	河流
river basin	流域
river bed	河床
river branching bay	河口湾
river landform	河流地貌
river mouth bar	河口沙坝
river terrace	河流阶地
riverbed placer	河床砂矿
river-dominated delta	河控三角洲

RMM	残留磁性测量
rock	岩石
rock anomaly	岩石异常
rock bench	海蚀阶地
rock crystal	水晶
rock province	岩省
rock salt	石盐
rockfall	崩塌
rock-forming element	造岩元素
rock-forming mineral	造岩矿物
rocking stone	摇摆石
rock-ore formation time interval	矿-岩时差
rocky desert	岩漠
roofing slate	瓦板岩
roggan(rocking stone)	摇摆石
rollover anticline structure	滚动背斜构造
Romania	罗马尼亚
root clay	根土岩
root marks	植物根痕
ropy lava	绳状熔岩
roscoelite	钒云母
rose color	玫瑰[红]色,淡红色
rose pink	淡粉红色
rose quartz	蔷薇石英
rosiness	玫瑰[红]色,淡红色
Rössing-type uranium deposit	勒辛型铀矿床
rosy	玫瑰[红]色,淡红色
rotation axis	旋转轴
rotation method	旋转法
rotational ellipsoid	旋转椭球体
roto-inversion	旋转反伸
roto-reflection	旋转反映
rough lead method	粗铅法

rounded	磨圆的,圆形的
roundness	圆度,磨圆度
row	行列
RR	反射折射仪
RS	拉曼光谱
ruarsite	硫砷钌矿
rubidium deposit	铷矿床
Rubidium(Rb)	铷
ruby	红宝石
ruby and sapphire deposit	红蓝宝石矿床
rufous	红褐色,赤褐色,红褐色
runoff	径流
russet	红褐色,赤褐色,红褐色
russety	红褐色,赤褐色,红褐色
Russia	俄罗斯
rust yellow	浅锈黄
rusty	铁锈色的
Ruthenium(Ru)	钌
rutile	金红石
Rwanda	卢旺达
RXMAM	若特维尔德 X 射线矿物含量分析方法(?)

S

S	熵
S wave	剪切波,横波(S 波)
sabkha	萨勃哈,潮上滩,盐沼,盐滩
sabkha metallogenic model	萨布哈成矿模式
sabkha salt model	萨布哈成盐模式
saccharoidal marble	糖粒状大理岩
saddle	鞍部
saddle reef	鞍状矿脉

saddle vein	鞍状矿脉
saddle-shaped	马鞍状
SAED	选区电子衍射
safflower yellow	红花黄
saffron [yellow]	橘黄色
Saint Helena and Ascension etc.	圣赫勒拿岛和阿森松岛等(英)
Saint Lucia	圣卢西亚
Saint Pierre and Miquelon	圣皮埃尔和密克隆群岛(法)
Saint Vincent and The Grenadines	圣文森特和格林纳丁斯
SALI	激光离子化表面分析
salic mineral	硅铝矿物
saline lake deposit	盐湖矿床
salinity	盐分，盐度
salinization	盐渍化
salite	次透辉石
salmon pink	橙红
salt	盐
salt deposit	盐类矿床
salt dome	盐丘
salt dome trap[structure]	盐丘圈闭[构造]
salt flat	盐滩(盐坪)
salt lake	盐湖
saltation [transport]	跃移[作用],跳移
saltation load	跳跃负载,跳跃总体
salt-lake-type boron deposit	盐湖型硼矿床
salt-lake-type lithium deposit	盐湖型锂矿床
Samarium(Sm)	钐
San Marino	圣马力诺
sand bar	沙坝
sand barrier	沙堤
sand beach	沙滩
sand coast	沙质海岸
sand mound	沙堆

sand spit	砂咀(沙嘴)
sandstone	砂岩
sandstone type U-ore	砂岩型铀矿
sandstorm	沙尘暴
sandwiched area of double fracture zone	双断裂夹持区
sandy beige	浅褐色
sandy desert	沙漠
sandy slate	砂质板岩
sanidine	透长石
sanidinite	透长岩
sanidinite facies	透长岩相
Sao Tome and Principe	圣多美和普林西比
sap green	暗绿色
sapanthracite	腐泥无烟煤
sapanthracon	腐泥烟煤
sapphire	蓝宝石
sapphire	天蓝
sapphirine	假蓝宝石
sappire	蓝宝石
saprocol	硬腐泥
saprocollite	胶泥煤
saprodite	腐泥褐煤
saprofication	腐泥化作用
sapropel	腐泥
sapropelite	腐泥煤
Saskatchewan	萨斯喀彻温省(加)
satellite	卫星
saturate	深色的
saturated	深色的
saturated rock	饱和岩
saturation	饱和,饱和度,色的纯度
Saudi Arabia	沙特阿拉伯
saussuritization	糟化,钠黝帘石化

saxe blue	萨克森蓝，水蓝，浅蓝色
scale	比例尺
scale model	比例模拟，尺度模拟
scalenohedron	偏三角面体
scaly	鳞片状的
Scandium(Sc)	钪
Scanning Electron Microscopy (SEM)	扫描电子显微镜，扫描电镜
Scanning Force Microscopy (SFM)	扫描力显微镜
Scanning Probe Microscopy (SPM)	扫描探针显微镜
Scanning Proton-Microprobe Analysis (SPMA)	扫描质子探针分析
Scanning Tunneling Microscopy (STM)	扫描隧道显微镜
Scanning Tunneling Spectroscopy (STS)	扫描隧道光谱
scapolite	方柱石
scapolitization	方柱石化
SCAR	南极考察科学委员会
scarcity of resources	资源的稀缺性
scarlet	鲜红，猩红，深红
scheelite	白钨矿
scheelite skarn deposit	矽卡岩白钨矿矿床
schiller	闪光，变彩
schist	片岩
schistic	片状的
schistose structure	片状构造
schistosity	片理
Schoenflies symbol	申弗利斯符号(圣弗利斯符号)
schorl	黑电气石
Schreinemakers' rule	施若内马克思法则
Schreinemakers' bundle	施若内马克思束
Scientific Committee on Antarctic Research (SCAR)	南极考察科学委员会
Scientific Committee on Oceanic Research (SCOR)	海洋研究科学委员会

sclerodermite	灰质簇,刺,板
sclerotinite	菌类体
scolecite	钙沸石
SCOR	海洋研究科学委员会
scoria cone	火山渣锥
scorodite	臭葱石
scour marks	冲刷痕
scour structure	冲刷构造
screw distance	螺距
screw rotation axis	螺旋轴
SCXRD	单晶 X 射线衍射
sea arch	海穹
sea cave	海蚀凹槽
sea cliff	海蚀崖
sea current	海流,洋流
sea deposit	海积作用
sea deposit landform	海积地貌
sea erosion landform	海蚀地貌
sea floor spreading	海底扩张
sea green	海绿色
sea wave	海浪
sea-beach placer	海滨砂矿
seal	密封,密封墙,封闭
sea-level change	海平面变化
sea-level pressure	海平面气压
seam	矿层
seashore	海滨
seaside	海滨
seasonal aspect of biome	群落的季相
second order transition	二级相变
secondary anomaly	次生异常
secondary building units	次级结构单元
Secondary Electron Imaging (SEI)	二次电子像

secondary enriched zone	次生富集带
secondary enriched zone of sulfide deposits	硫化物矿床次生富集带
secondary enrichment	次生富集作用
secondary enrichment of sulfide deposits	硫化物矿床次生富集作用
secondary environment	次生环境
Secondary High-Mass Resolution Ion Microprobe (SHRIMP)	二次高质量共振离子探针
secondary inclusion	次生包体
secondary joint	次生节理
secondary mineral	次生矿物
secondary porosity	次生孔隙度
secondary production	次级生产(加工工业)
secondary quartzite	次生石英岩
secondary twin	次生双晶
Secondary-Ion Mass-Spectrometry (SIMS)	二次离子质谱仪
secretion	分泌体
sedimentary model of coal	煤沉积模式
sedimentary systems	沉积体系
sedimentary basins	沉积盆地
sedimentary cycle in coal-bearing series	含煤岩系旋回结构
sedimentary deposit	沉积矿床
sedimentary environment	沉积环境
sedimentary exhalative deposit	沉积喷流矿床
sedimentary facies	沉积相
sedimentary iron deposit	沉积铁矿床
sedimentary rock	沉积岩
sedimentation	沉积[作用]
sedimentology	沉积学
sedimento-metamorphic deposit	沉积变质矿床
sedimento-metamorphic manganese deposit	沉积变质锰矿床
sedimento-metamorphic phosphate deposit	沉积变质磷矿床
seepage flow	渗流
segregation	分凝作用

SEI	二次电子像
seism	地震
seismic focus	震源
seismic sea wave	海啸
seismic sequence	地震序列
seismic source	震源
seismic source depth	震源深度
seismic wave	地震波
seismic zone	地震带,贝尼奥夫带
seismic zoning map	地震区划图
seismism	地震作用
seismological	地震的
Selected-Area Electron Diffraction (SAED)	选区电子衍射
selenium deposit	硒矿床
Selenium(Se)	硒
selfconfinement	自限性
self-diffusion	自扩散作用
SEM	扫描电子显微镜,扫描电镜
semibright coal	半亮煤
semidull coal	半暗煤
semifusinite	半丝质体
semi-pelite	半泥质岩
semi-pelitic phyllite	半泥质千枚岩
semi-pelitic schist	半泥质片岩
Senegal	塞内加尔
sense of displacement	位移方向
Sensitive High-Mass Resolution Ion Microprobe (SHRIMP)	灵敏高质量分辨离子探针
separation	断距
sepiolite	海泡石
sepiolite clay deposit	海泡石粘土矿床
sepiolite deposit	海泡石矿床
sequence	生成顺序

sequential extraction	顺序提取
seriate porphyritic texture	连续不等粒斑状结构
sericite	绢云母
sericitization	绢云母化
Series	统
serpentine	蛇纹石
serpentine asbestos	蛇纹石石棉
serpentine-talc schist	蛇纹石-滑石片岩
serpentinite	蛇纹岩
serpentinization	蛇纹石化
settling velocity	沉降速度
SEXAFS	表面扩展 X 射线吸收精细结构光谱
Seychelles	塞舌尔
SFM	扫描力显微镜
SG	特殊重力测试
shading yellow	暗黄色
shagreen surface	糙面
shale	页岩,泥板岩
shallow burial diagenesis	浅埋成岩作用
shallow intrusion	浅成侵入体
shallow sea	浅海
shallow-focus-earthquake	浅源地震
sheaf texture	束状结构
shear fold	剪切褶皱
shear joint	剪节理
shear stress	剪应力
shear wave(S wave)	剪切波(S 波)
sheath fold	鞘褶皱
sheepback rock	羊背石
sheet	片,图幅,岩席
sheet silicate	层状硅酸盐
sheeted-zone veins	重膜状矿脉

shell rule	壳层规则
shield volcano	盾[形]火山
shocking pink	鲜粉红色
shoestring sandstone	鞋带状砂岩
shore	滨海
short-range forecast	短期预报
Short-Wave Infrared Spectroscopy (SWIS)	短波红外光谱
SHRIMP	二次高质量共振离子探针
SHRIMP	灵敏高质量分辨离子探针
SHRIMP	单颗粒锆石离子探针质谱法
SHRIMP(The zircons were analyzed by a specially built high resolution ion microprobe at the Australian National University nicknamed 'SHRIMP')	单颗粒锆石离子探针质谱法
sial	硅铝层
Siderian	成铁系
siderite	菱铁矿
sideronitic texture	海绵陨铁结构
siderophile element	亲铁元素(过渡型元素)
siderophyllite	铁叶云母
siderophyre	古铜-鳞英石铁陨石
sienna	赭[石]色
Sierra Leone	塞拉里昂
sieve texture	筛状结构
significant anomaly	有意义异常
Sikkim	锡金
Silesian-type lead-zinc deposit	西里西亚式铅锌矿床
silic-alkali index	硅碱指数
silicate	硅酸盐
silicic	酸性的,硅质
siliciclastic	硅质碎屑的,硅屑的
silicification	硅化
silicified zone type U-ore	硅化带型铀矿

Silicon(Si)	硅
silky luster	丝绢光泽
sill	岩床
sillimanite	夕线石
sillimanite gneiss	夕线石片麻岩
silt	粉砂
siltstone	粉砂岩
Silurian	志留系
Silurian Period(System)	志留纪(系)
silver	银[白]色
silver deposit	银矿床
Silver(Ag)	银
sima	硅镁层
similar fold	相似褶皱
simple decarbonation reaction	简单的脱碳酸反应
simple dehydration reaction	简单的脱水反应
SIMS	二次离子质谱仪
simulation	模拟
simulation experiment	模拟实验
Singapore	新加坡
single chain silicate	单链状硅酸盐
single form	单形
single form symbol	单形符号
single stage system	单阶段体系
Single-Crystal Refinement X-ray Analysis (SREF-XRA)	单晶精细 X 射线分析
Single-Crystal X-ray Diffraction (SCXRD)	单晶 X 射线衍射
singularity	奇点
Sinian Period(System)	震旦纪(系)
SIRMS	稳定同位素比率质谱仪
situ	原地,就地,在原处
skarn	矽卡岩

skarn copper deposit	矽卡岩型铜矿床
skarn iron deposit	矽卡岩型铁矿床
skarn mineral deposit	矽卡岩矿床
skarn molybdenum deposit	矽卡岩型钼矿床
skeletal	骸晶的，骨骼的，骨架的
skeletal morphology	骸晶形貌
skeleton crystal	骸晶
skeleton texture	骸晶结构
sketch	草图，素描
sky blue	天蓝
skyey	天蓝
slacktip	崩塌
SLAP(standard light Antarctic precipitation)	SLAP 标准(国际原子能机构推荐的第二个氢、氧同位素标准，$\delta D=-428‰$)
slate	板岩
slaty cleavage	板劈理
slaty structure	板状构造
slide	载玻片
sliding reaction	滑动反应
slip	滑距
slope of isochron	等时线斜率
slope-wash iron deposit	坡积铁矿床
Slovakia	斯洛伐克
Slovenia	斯洛文尼亚
slump	滑塌
SM	立体显微镜，双目显微镜
smectite	蒙脱石
smithsonite	菱锌矿
Sm-Nd dating	钐-钕计时
smoky gray	烟灰色
smoky quartz	烟水晶

SMOW(standard mean ocean water)	SMOW 标准(标准平均大洋水，氢、氧同位素标准)
smut	煤垩
snow	雪
snow line	雪线
snowball texture	雪球结构
snowy white	雪白色
snuff color	黄褐色
soapstone	块滑石，寿山石
sociality of resources	资源的社会性
soda niter	钠硝石
soda niter deposit	钠硝石矿床(智利硝石矿床)
sodalite	方钠石
sodic metasomatism	钠交代[作用]
sodic-metasomatism type uranium deposit	钠交代型铀矿
Sodium(Na)	钠
soil	土壤
soil anomaly	土壤异常
soil classification	土壤分类
soil colloid	土壤胶体
soil erosion	水土流失，土壤侵蚀
soil fertility	土壤肥力
soil porosity	土壤孔隙度
soil profile	土壤剖面
solar constant	太阳常数
solar system	太阳系
solar system abundance of element	太阳系元素丰度
solar wind	太阳风
sole marks	底痕
solid solution	固溶体
solid state chemistry	固态化学
solid tide	固体潮
solid-solid reaction	固-固反应

solidus	固相线
Solomon Islands	所罗门群岛
solubility	溶解度
solubility product	溶度积
solum	土体
solution	溶解
solution depression	溶蚀洼地(溶洼)
solution-cavity filling	溶洞充填
solvus	固溶体分解曲线
solvus exsolution curve	固溶体分解曲线
Somalia	索马里
sorosilicate	双岛状硅酸盐
sorting	分选,分选作用
source bed	矿源层,烃源层(生油层)
source rock	矿源岩,烃源岩(生油岩)
source rock evaluation	生油岩评价
South Africa	南非
South Carolina	南卡罗来纳州(美)
South Dakota	南达科他州(美)
space chemistry	空间化学
space geology	宇宙地质学
space group	空间群
spaced cleavage	间隔劈理
Spain	西班牙
spallogenic nuclide	散裂成因核素
sparry	亮晶
sparry cement	亮晶胶结物
spatter cone	溅落锥
spatter rampart	溅落堤
special form	特殊[单]形
species	种(矿物分类基本单位)
species extinct	物种灭绝
specific gravity	比重

Specific Gravity Determinations (SG)	特殊重力测试
specimen	标本，样品，样本
speculative resource	理论资源
spessartine	锰铝榴石
spessartite	锰铝石榴子石,闪斜煌岩
sphalerite	闪锌矿
sphene	榍石
sphenoid	楔形
sphere	球形
spherical coordinate	球面坐标
spherical projection	球面投影
spheroidal weathering	球形风化
spherulitic	球粒状
spike	稀释剂
spilite	细碧岩
spindle stage	旋转针台
spinel	尖晶石
spinel deposit	尖晶石矿床
spinifex texture	鬣刺结构
spiral arm	悬臂
spiral galaxy	漩涡星系
spiral steps	螺旋生长台阶
S-planes	S 面
splitting of root	根劈作用
SPM	扫描探针显微镜
SPMA	扫描质子探针分析
spodumene	锂辉石
spodumene-lepidolite-pegmatite deposit	伟晶岩锂矿床(锂辉石-锂云母伟晶岩矿床)
sponge	海绵，海绵体，海绵状物
spongolite	海绵岩
spontaneous fission	自发裂变
sporinite	孢子体

spotted slate	斑点板岩
spotted structure	斑点[状]构造
spurrite	灰硅钙石
square grid	正方形格子
squared paper	方格纸
SREF-XRA	单晶精细 X 射线分析
Sri Lanka	斯里兰卡
Sr-O system	锶-氧体系
S-surfaces	S 面
stability	稳定性
stable isotope geochemistry	稳定同位素地球化学
Stable Isotope Ratio Mass Spectrometry (SIRMS)	稳定同位素比率质谱仪
stable state	稳[定状]态
stack	海蚀柱
stage	期,阶
stage of mineralization	成矿阶段
stalactite	石钟乳
stalactitic aggregate	钟乳状集合体
stalacto-stalagmite	石柱
stalagmite	石笋
stalagmitic	石笋状
star	恒星
star dune	星状沙丘
state of strain	应变状态
state of stress	应力状态
Statherian	固结系
static recrystalizaion	静态重结晶
statistical petrology	统计岩石学
staurolite	十字石
steady-state geothermal gradient	稳态地热梯度
streaky gneiss	条痕状片麻岩
Stenian	狭带系

step fault 阶梯状断层
stepwise heating 阶段加温
stereogram 三维图形
stereographic projection 极射赤平投影
Stereoscopic Microscopy(SM) 立体显微镜,双目显微镜
stibarsen 砷锑矿
stibnite 辉锑矿
stilbite 辉沸石
stilpnomelane 黑硬绿泥石
stipoverite 斯石英
STM 扫描隧道显微镜
stock 岩株
stock resource 储存性资源
stockwork 网脉
stockwork and disseminated molybdenum deposit 细脉浸染型钼矿床
stockwork ore zone 网脉状矿石带
stoichiometry 化学计量学
stone blue 灰蓝色
stone yellow 土黄,淡黄
stony meteorite 石陨石
stony-iron meteorite 石铁陨石
storm current 风暴流
straight extinction 直消光,平行消光
strain ellipsoid 应变椭球
strain field 应变场
stratabound lead-zinc deposit in carbonate rocks 碳酸盐岩层控铅锌矿床
strata-bound mineral deposit 层控矿床
strategy 对策
stratification 层理,层化
stratiform [mineral] deposit 层状矿床
stratiform copper deposit 层状铜矿床

stratigraphic column	地层柱
stratigraphic correlation	地层对比
stratigraphic geochemistry	地层地球化学
stratigraphic section	地层剖面
stratigraphic trap	地层圈闭
stratigraphic well	基准井,参数井
stratigraphical sampling	分域采样
stratigraphy	地层学
stratosphere	平流层
stratovolcano	层[状]火山
stratum	地层
streak	条痕
streaky migmatite	条痕状混合岩
stream sediment survey	水系沉积物测量
stress	应力
stress difference	应力差
stress field	应力场
stress mineral	应力矿物
stretching lineation	拉伸线理
striae	晶面条纹,生长条纹
striation	晶面条纹,生长条纹
strike	走向
strike fault	走向断层
strike-slip fault	平移断层,走[向]滑[动]断层
stringer lode	网脉
striped magnetite quartzite	条带状磁铁石英岩
striped migmatite	条带状混合岩
stromatactis	平底晶洞构造,层状孔洞构造,叠层生物孔层构造
stromatic structure	层状构造,带状构造
stromatite	层(条带)状混合岩
stromatolite	叠层石
stromatolitic limestone	叠层石灰岩

stromatolitic structure	叠层构造
stromatoporoid	层孔虫类
strong yellowish pink	深黄红色
Strontium(Sr)	锶
structural analysis	构造分析
structural data	构造资料
structural feature	构造形迹
structural geology	构造地质学
structural landform	构造地貌
structural map	构造图
structural mountain	构造山系
structural petrology	构造岩石学
structural trap	构造圈闭
structure	构造
structure contour	构造等值(高)线
structure factor	结构因子
structure of ores	矿石构造
STS	扫描隧道光谱
study of ore deposit	矿床学
stumpy	短柱状
stylolite	缝合线(面)
subangular	次棱角状的
subaqueous	水下的
subaqueous gliding	水下滑移[作用]
subaqueous slump	水下滑移[作用]
subcontinental lithosphere	陆下岩石圈
subduction	俯冲
subduction zone	消减带
suberinite	木栓质体
subface	底面
subgrain	亚颗粒
subgroup	子群
sublimation	升华作用

submarine eruption	海底喷发
submarine exhalative process	海底喷流作用
submarine heat flow	海底热流
submarine hot spring	海底热泉
submarine metallization	海底成矿作用
submarine metallization at East Pacific Rise	东太平洋洋隆成矿作用
submarine metallization in the Red Sea	红海海底成矿作用
submetallic luster	半金属光泽
subrounded	次圆状的,半磨圆的
subsequent river	后成河
subsidence	沉降
subspecies	亚种(矿物分类单位)
substitution	置换
substrate	基体
substratum	基体
sub-surface structure	地下构造
subtidal zone	潮下带
subtle trap	隐蔽圈闭
subtropical high pressure	副热带高压
subvolcanic facies	潜火山相
subvolcanic rock	潜火山岩
succession of biome	群落的演退
sucrosic dolomite	糖粒状白云岩
Sudan	苏丹
Sudbury copper-nickel deposit	萨德伯里铜镍矿床
sudoite	铝绿泥石
Suess effect	苏斯效应
suevite	冲击角砾岩(陨击角砾岩,撞击角砾岩)
suffusion	潜蚀[作用]
sulfur cycle	硫循环
sulfophile element	亲铜元素,亲硫元素
sulfur-containing organic compound	有机含硫化合物

sulphate	硫酸盐
sulphide	硫化物
sulphur	自然硫
sulphur deposit	硫矿床
Sulphur(S)	铳
sun system	太阳系
sunspot cycle	黑子周期
sunstone	日光石
suolunite	索伦石
super plume	超地幔柱
super structure	超结构
super-cell	超晶胞
super-critical state	超临界状态
superdense hypothesis	超密说
superface	顶面
superficial ooid	表鲕
supergene mineral	表生矿物
supergenesis	表生作用
supergroup	母群
superimposed anomaly	叠加异常
superlattice	超晶格
supernova explosion	超新星爆发
superposed fold	叠加褶皱
supersaturation	过饱和度
supporting style	支撑类型
supracritical ore fluid	超临界含矿流体
supracrustal rock	表壳岩
surf zone	激浪带
Surface Analysis By Laser Ionization (SALI)	激光离子化表面分析
Surface-Extended X-ray Absorption Fine-Structure Spectroscopy (SEXAFS)	表面扩展 X 射线吸收精细结构光谱

surge	风暴潮
Suriname	苏里南
suspend	悬浮
suspended load	悬浮负载,悬浮总体
suspended particle	悬浮粒
suspended water	包气带水
suspension	悬运
suspension transport	悬移[作用]
sustainable capacity	持续能力
sustainable development	可持续发展
sustainable use	可持续利用
sustainable use of natural resources	自然资源的可持续利用
suture [line]	缝合[线]
suture zone	缝合带
swamp	沼泽
swampy land	湿地
swarthy	浅黑的
swash cross-bedding	冲洗交错层理
Swaziland	斯威士兰
Sweden	瑞典
SWIS	短波红外光谱
Switzerland	瑞士
SXFM	同步辐射 X 射线荧光探针分析
SXRT	同步辐射 X 射线形貌分析
syenite	正长岩
sylvine	钾盐
sylvine deposit	钾盐矿床
symmetric extinction	对称消光
symmetric operation matrix	对称变换矩阵
symmetrical banded structure	对称带状构造
symmetrical fold	对称褶皱
symmetrical ripples	对称波痕
symmetry	对称

symmetry axis	对称轴
symmetry operation	对称操作
symmetry plane	对称面
symmetry element	对称要素
symplectite	后成合晶
Synchrotron X-ray Fluorescence microprobe Analysis(SXFM)	同步辐射 X 射线荧光探针分析
Synchrotron X-ray Topography(SXRT)	同步辐射 X 射线形貌分析
syncline	向斜
synclinorium	复向斜
syndepositional	同生沉积的,沉积同期的,与沉积同时的
synform	向形
syngenetic anomaly	同生异常
syngenetic deposit	同生矿床
syngenetic hydrothermal process	同生水热作用
syngenetic ore-forming process	同生成矿作用
syngenetic ore-forming theory	同生成矿说
syntectonic blast	同构造变晶
synthesize ooid	复鲕
synthetic mineral	人工矿物,合成矿物
Syria	叙利亚
System	系
system components	系统组分
szaibelyite	硼镁石
szaibelyite deposit	硼镁石矿床

T

tabular	[厚]板状
tabular cross-bedding	板状交错层理
tabular U-ore body	板状铀矿体
TAC	季节性吸收系数
taconite	铁英岩

taenite	镍纹石
Tajikistan	塔吉克斯坦
talc	滑石
talc deposit	滑石矿床
talc schist	滑石片岩
talus	倒石堆
tan	茶[褐]色,棕褐色,咖啡色
Tantalum(Ta)	钽
Tanzania	坦桑尼亚
tapered	锥形的,尖灭的
Tapping Mode Atomic Force Microscopy (TMAFM)	轻敲模式原子力显微镜
tar sand	沥青砂
taramite	绿闪石
target selection	靶区优选
tarnish	锖色
tartan structure	格子结构
tartan twin	格子双晶
tartan twinning	格子双晶
taupe	灰褐色
tawny	茶[褐]色,棕褐色,咖啡色
taxitic structure	斑杂构造
Technetium(Tc)	锝
technological petrology	工艺岩石学
tectonic basin	构造盆地
tectonic breccia	构造角砾岩
tectonic conglomerate	构造砾岩
tectonic cycle	构造旋回
tectonic earthquake	构造地震
tectonic joint	构造节理
tectonic lake	构造湖
tectonic landform	构造地貌
tectonic overpressure	构造超压

tectonic regime	构造体制
tectonic setting	构造背景
tectonics	大地构造学,构造学
tectonite	构造岩
tectonization	构造运动
tectono-geochemistry	构造地球化学
tectosilicate	架状硅酸盐
tektite	玻陨石,熔融石
telescoped deposit	叠套矿床,叠生矿床
telescoping	叠生作用,叠套作用
telethermal deposit	远成热液矿床
telinite	结构镜质体
tellurium deposit	碲矿床
Tellurium(Te)	碲
TEM	透射电子显微镜
temperature gradient	温度梯度
temperature inversion	逆温
temporal absorption coefficient(TAC)	季节性吸收系数
tenacity	韧性,粘滞性
tengchongite	腾冲铀矿
Tennessee	田纳西州(美)
tenor of ore	矿石品位
tension joint	张节理
tensor	张量
tentative isograd	准等变线
tephra	火山碎屑,爆发岩屑
tephra cone	火山碎屑锥
tephra fall	火山碎屑降落
tephra flow	火山碎屑流
tephrite	碱玄岩
Terbium(Tb)	铽
terminal reaction	终端反应
ternary eutectic	三元共晶

ternary system	三元[体]系
terrace	阶地
terrain	地体
terrestrial	地表的
terrestrial age	陨石落地年龄
terrestrial ecosystem	地表生态系统
terrestrial meridian	地球子午圈
terrestrial planet	类地行星
terrigenous	陆源的,陆成的,陆生的
terrigenous clastic rock	陆源碎屑岩
terrigenous clastic sediment	陆源碎屑沉积物
terrigenous detrital	陆源碎屑
terrigenous detrital sediment	陆源碎屑物
Tertiary Period(System)	第三纪(系)
test plate	试板
Tethyan metallogenic belt	特堤斯成矿带
Tethys	特提斯(海)
tetraauricupride	四方铜金矿
tetragonal bipyramid	四方双锥
tetragonal disphenoid	四方双锥(楔)
tetragonal prism	四方柱
tetragonal pyramid	四方单锥
tetragonal system	四方晶系
tetragonal tetrahedron	四方四面体
tetragonal trapezohedron	四方偏方三八面体
tetragonal trisoctahedron	四角三八面体
tetrahedrite	黝铜矿
tetrahedron	四面体
tetrahexahedron	四六面体
tetrakishexahedron	四六面体
tetrakisoctahedron	四角三八面体
tetrakistetrahedron	四角三四面体
Texas	得克萨斯州(美)

textural maturity	结构成熟度
texture	结构
texture of ores	矿石结构
texture reverse	结构退变,结构倒转
TGA	热重分析
Th/U ratio	钍铀比
Thailand	泰国
thallium deposit	铊矿床
Thallium(Tl)	铊
thallus	原植物,菌体
The Commonwealth of Dominica	多米尼加联邦
The Commonwealth of the Northern Mariana Islands	北马里亚纳群岛自由联邦(美)
The Democratic People's Republic of Korea	朝鲜民主主义人民共和国
The Dominican Republic	多米尼加共和国
The Federated States of Micronesia	密克罗尼西亚联邦
The Federation of Saint Kitts and Nevis	圣基茨和尼维斯联邦
the Milky Way	银河[系]
The Netherlands Antilles	荷属安的列斯
the origine of life	生命起源
The Republic of Belau	贝劳共和国
The United Arab Emirates	阿拉伯联合酋长国
the visible universe	可见宇宙
theory of concentration of metallogenesis	矿化集中(金属)区学说
theory of nucleosynthesis	元素合成理论
theralite	霞斜岩
thermal and sound insulating materials	隔热隔音材料
thermal conductivity	热传导率
thermal disturbance	热扰动
thermal electricity	热电性
thermal expansion	热膨胀
Thermal Gravimetric Analysis(TGA)	热重分析

Thermal Ionization Mass Spectroscopy (TIMS)	热离子化质谱仪
thermal peak	热峰
thermal peak conditions	热峰条件
thermal relaxation	热松弛
thermal reworked type uranium deposit	热改造型铀矿床
thermal ripening	热成熟
thermal transformation	热转变
thermal weathering	温差风化
thermal-peak isobars	热峰等压线
thermal-peak isotherms	热峰等温线
thermal-peak pressure	热峰压力
thermal-peak temperature	热峰温度
thermobarogeochemistry	温压地球化学,热压地球化学
thermodynamic	热力学的
thermodynamics	热动力学
thermoelectricity	热电性
thermoluminescence	热发光,热致发光
thermometamorphic rock	热变质岩
thermometamorphism	热变质作用
thermo-simulation of organic matter	有机质热模拟
thickness of stratum	岩层厚度
thin section	薄片
thin section method	薄片鉴定法
third order pink	三级红
tholeiite	拉斑玄武岩
tholeiitic series	拉斑系列
tholeiitic texture	间隐结构
thomsonite	杆沸石
thorium deposit	钍矿床
Thorium(Th)	钍
three dimensional zonality	三维地带性
three dimentional diagram	三维图,立体图

threshold	异常下限
threshold of oil generation	生油门限
thrust system	逆冲断层系
Thulium(Tm)	铥
thumb blue	靛蓝,紫蓝
thuringite	鳞绿泥石
tidal flat	潮坪,湖滩
tidal range	潮差
tide	潮汐
tie line	连岛沙坝
tieline-switching reaction	连线转换反应
till	冰碛,冰碛物
till landform	冰碛地貌
tilleyite	粒硅钙石
time temperature index(TTI)	时间温度指数
TIMS	热离子化质谱仪
tin deposit	锡矿床
tin placer	砂锡矿床
Tin(Sn)	锡
tinge	淡色
titanite	榍石
titanium deposit	钛矿床
Titanium(Ti)	钛
TMAFM	轻敲模式原子力显微镜
Togo	多哥
Tokelau Islands	托克劳群岛(新)
tonalite	英云闪长岩
tonalitic gneiss	英云闪长质片麻岩
Tonga	汤加
tongbaite	桐柏矿
Tonian	拉伸系
tool marks	工具痕,流痕,水流蚀痕
topaz	黄玉

topochemical	局部化学的
topographic isotopic fractionation effect	地形同位素效应
topological	拓扑学的,地志学的
topology	拓扑学,地志学
torbanite	藻烛煤
torbernite	铜铀云母
tornado	龙卷风
torrent flow	急流
Torsion Balance	扭力天平
tourmaline	电气石
tourmalinization	电气石化
trace element	微量元素,痕量元素(微迹元素)
trachyandesite	粗安岩
trachybasalt	粗玄岩
trachyte	粗面岩
trachytic texture	粗面结构
tracks	足迹,虫迹
traction	推移[作用]
traction [load]	牵引负载(牵引总体)
tractive current	牵引流
trade wind	信风
trails	行迹,爬迹
transient geothermal gradient	瞬时地热梯度
transformation twin	转变双晶
tranquil flow	缓流
tranquillityite	静海石
transform boundary	转换断层型边界
transform fault	转换断层
transformation of environmental substance	环境物质转化
transgression	海进
transition bond	过渡型键
transitional element	过渡型元素(亲铁元素)
transitional support	过渡性支撑

transitional zone	过渡带
translation axis	平移轴
translation gliding	直线滑移
translucent	半透明
translucent opaque	不透明
Transmission Electron Microscopy(TEM)	透射电子显微镜
trans-opaque	半透明
transparency	透明度
transparent	透明
transparent mineral	透明矿物
transport of environmental substance	环境物质迁移
transportation	搬运[作用]
transversal dune	横向沙丘
transversal zoning	横向分带
transverse wave(S wave)	横波(S 波)
trap	暗色岩
trapezohedron	偏方面体,四角三八面体
travel time	走时
travertine	石灰华
tremolite	透闪石
trench	海沟
triakisoctahedron	三角三八面体
triakistetrahedron	三角三四面体
Triassic	三叠系
Triassic Period(System)	三叠纪(系)
triboluminescence	摩檫发光
triclinic system	三斜晶系
third order transition	三级相变
tridymite	鳞石英
trigonal bipyramid	三方双锥
trigonal prism	三方柱
trigonal pyramid	三方单锥
trigonal system	三方晶系,菱面体晶系

trigonal trapezohedron	三方偏方面体
Trimacerite	微三合煤
trimorphism	同质三象
Trinidad and Tobago	特立尼达和多巴哥
triple junction	三联点,三向联结构造(三叉点)
triple point	三向联结构造(三叉点)
troctolite	橄长岩
troilite	陨硫铁
trona	天然碱
trona deposit	天然碱矿床
Trondhjemite	奥长花岗岩
trondhjemitic gneiss	奥长花岗质片麻岩
trophic level	营养级
tropical cyclone	热带气旋
troposphere	对流层
trough	槽
trough cross-bedding	槽状交错层理
true isograd	真等变线
true ooid	真鲕
tschermakite	钙镁闪石
tsilaisit	钠锰电气石
tsunami	海啸
TTI	时间温度指数
tuff	凝灰岩
tuff texture	凝灰结构
tungstate	钨酸盐
tungsten deposit	钨矿床
Tungsten(W)	钨
tungsten-antimony-gold-quartz vein	钨锑金石英脉
Tunisia	突尼斯
turbidite	浊积岩,浊流层
turbidity	混浊性
turbidity current	浊流

turbulent flow	湍流,紊流
Turkey	土耳其
Turkmenistan	土库曼斯坦
Turks and Caicos Islands	特克斯和凯科斯群岛(英)
turmeric yellow	姜黄
turquoise	绿松石
turquoise blue	绿松石色,土耳其玉色,青绿色
turquoise deposit	绿松石矿床
Tuvalu	图瓦卢
tweed texture	花呢结构
twin axis(twinning axis)	双晶轴
twin center	双晶中心
twin crystal	双晶
twin element	双晶要素
twin lamellae	双晶片
twin law	双晶律
twin partition line	双晶缝合线
twinkling	闪突起
twinning	双晶
twinning plane	双晶面
twinning striation	双晶条纹
two circle goniometer	双圈测角仪
two stage system	二阶段体系
two-dimensional nucleus	二维核
two-feldspar gneiss	二长片麻岩
type	类(矿物分类单位)
type of metamorphism	变质作用类型
typhoon	台风
typic structure	典型结构
typochemical element	标型元素
typomorphic characteristic	标型特征
typomorphic mineral	标型矿物
typomorphism	标型性

tyuyamunite	钒钙铀矿

U

U-[shaped] valley	U 形谷
U-[Th]-Pb dating	铀-[钍]-铅计时
Uganda	乌干达
UHP	超高压
UHV STM	紫外高真空扫描隧道显微镜
UI	超声波干扰(干涉)测量
Ukraine	乌克兰
ultimately recoverable resource	最终可采资源
ultraacidic rock	超酸性岩
ultrabasic	超基性
ultrabasic gneiss	超基性片麻岩
ultrabasic rock	超基性岩
ultracataclasite	超碎裂岩
ultra-high pressure(UHP)	超高压
ultra-high-pressure metamorphic rock	超高压变质岩
ultramafic	超镁铁质
ultramafic rock	超镁铁[质]岩
ultramarine blue	群青,暗青绿色,深蓝
ultrametamorphic rock	超变质岩
ultrametamorphism	超变质[作用]
ultramicroscopic	超显微的
ultramylonite	超糜棱岩
Ultrasonic Interferometry(UI)	超声波干扰(干涉)测量
Ultraviolet High-Vacuum Scanning Tunneling Microscopy(UHV STM)	紫外高真空扫描隧道显微镜
Ultraviolet Laser Ablation Mass Spectroscopy(UV-LA-MS)	紫外激光磨蚀质谱仪
Ultraviolet Photoelectron Spectroscopy (UPS)	紫外光电子光谱
ulvospinel	钛尖晶石

umber	棕色,赭土
UMC	上地幔委员会
unconformity	不整合
unconformity trap	不整合圈闭
unconformity-vein type uranium deposit	不整合脉型铀矿
unconsolidated	未固结的
undation theory	波动说
under-development	欠发达
underground water	地下水,潜水
underlying surface	下垫面
underplating	底浸
underthrust fault	俯冲断层
undulating extinction	波状消光
undulose extinction	波状消光
UNESCO	联合国科教文组织
uneven fracture	参差状断口,粗糙断口
uniaxial crystal	一轴晶
uniformitarianism	均变论
unit cell	晶胞
United Kingdom	英国
United Nations Educational Scientific and Cultural Organization (UNESCO)	联合国科教文组织
United States of America	美利坚合众国
univariant curve	单变线
univariant equilibrium	单变平衡
universal stage	弗氏旋转台
universe	宇宙
unsaturated rock	不饱和岩
unstability	不稳定性
unstable state	不稳[定状]态
U-ore of chlorite type	绿泥石型铀矿
U-ore of explosion-breccia type	爆破角砾岩型铀矿
U-ore of microcrystalline quartz type	微晶石英型铀矿

Upper /Late Cambrian	上寒武统
Upper /Late Cretaceous	上白垩统
Upper /Late Devonian	上泥盆统
Upper /Late Jurassic	上侏罗统
Upper /Late Ordovician	上奥陶统
Upper /Late Triassic	上三叠统
upper hemisphere	上半球
Upper Mantle Committee (UMC)	上地幔委员会
upright fold	直立褶皱
UPS	紫外光电子光谱
upthrown side	上升盘
upwelling-current mineralization	上升洋流成矿作用
upwelling-current model of phosphate deposit	上升洋流磷矿成矿模式
U-Ra equilibrium	铀-镭平衡
uralitization	纤闪石化
uraniferous collophane	含铀胶磷矿
uraninite	晶质铀矿
uranium black	铀黑
uranium deposit	铀矿床
uranium geology	铀矿地质学
uranium mobilization	铀活化
uranium oxidized zone	铀氧化带
uranium productive massif	产铀岩体
uranium reduction zone	铀还原带
uranium roll	卷状铀矿体
uranium series dating	铀系计时
Uranium(U)	铀
uranium-bearing shale	含铀页岩
uranium-bearing silicalite	含铀硅岩
uranium-rich massif	富铀岩体
urtite	磷霞岩
Uruguay	乌拉圭

useful mineral	矿产
Utah	犹他州(美)
uvarovite	钙铬石榴子石
UV-LA-MS	紫外激光磨蚀质谱仪
UV-Visible Spectroscopy (UV-VS)	紫外可见光谱
UV-VS	紫外可见光谱
Uzbekistan	乌兹别克斯坦

V

vacancy	空缺,缺位
vadose silt	渗滤粉砂
vadose zone	渗流区
valence band	价带
valley	河谷
valley terrace	河谷阶地
vally placer	河谷砂矿
Van dee Waals' bond	范德华键
vanadate	钒酸盐
vanadic titanomagnetite deposit	钒钛磁铁矿矿床
vanadinite	钒铅矿
Vanadium(V)	钒
Vanuatu	瓦努阿图
vapor pressure	蒸汽压
variation diagram of element abundance	元素丰度变化图
variety	变种(矿物分类单位)
variole	球颗
variolite	球颗玄武岩
variolite texture	[玄武]球颗结构
variolitic texture	球颗结构
various form	变形
varve	纹泥
varve chronology	纹理年代学
varved clay	纹泥

vat blue	瓮蓝
Vatican City State	梵蒂冈
vector	矢量
vein	脉，岩脉，矿脉
vein material	脉体
vein structure	脉状构造
vein-type copper deposit	脉状铜矿床
veinlet-disseminated copper deposit	细脉浸染型铜矿床
Venetian blue	威尼斯蓝
Venezuela	委内瑞拉
venite	脉状混合岩
vent	喷出口
ventifact	风棱石
verdure	碧绿
vermeil	朱红
vermiculite	蛭石
vermiculite deposit	蛭石矿床
vermilion	朱红
Vermont	佛蒙特州（美）
Vernasky	魏尔纳斯基
vertical fold	倾竖褶皱
vertical movement	升降运动，垂直运动
vertical succession	垂向序列
vertical zoning	垂直分带
very coarse sandstone	极粗砂岩
very fine sandstone	极细砂岩
very high relief	很高突起
very low-grade metamorphic rock	很低级变质岩
vesicular	有气孔的，多孔状构造的
vesicular structure	气孔构造
vesignieite	钒钡铜矿
vesuvianite	符山石
VHN	维氏硬度计

Vicker's microhardness	维克斯显微硬度,维氏显微硬度
Vickers' Hardness Number(VHN)	维氏硬度计
Viet Nam	越南
violet	紫罗兰色,紫色
violet carmine	紫红
violet red	紫红
virescent	淡绿色
Virgin Islands of the United States	美属维尔京群岛
Virginia	弗吉尼亚州(美)
viridescence	嫩绿
viridescent	淡绿色
virtual inert components	有效惰性组分
viscosity	粘性,粘度,粘滞性
Visible and Near-Infrared Reflectance Spectroscopy (VNIRS)	可见与近红外反射光谱
vitrain	镜煤
vitreous	玻璃质
vitreous luster	玻璃光泽
vitric pyroclast	[火山]玻屑
vitrinertite	微镜惰煤
vitrinite	镜质组
vitrite	微镜煤
vitroclastic texture	玻璃质碎屑结构
vitrodetrinite	碎屑镜质体
vitroporphyric texture	玻基斑状结构
vivianite	蓝铁矿
vivid red	亮红色
VNIRS	可见与近红外反射光谱
vogesite	闪正煌岩
void	空隙
volatile elements	挥发性元素
volcanic activity	火山活动
volcanic apparatus	火山机构

volcanic arc	火山弧
volcanic arc rock	火山弧岩石
volcanic ash	火山灰
volcanic belt	火山带
volcanic bomb	火山弹
volcanic conduit	火山通道
volcanic cone	火山锥
volcanic cycle	火山旋回
volcanic dome	火山穹隆
volcanic earthquake	火山地震
volcanic edifice	火山机构,火山机体
volcanic facies	火山岩相
volcanic feeder	火山通道
volcanic glass	火山玻璃
volcanic landform	火山地貌
volcanic mudflow	火山泥[石]流
volcanic neck	火山颈
volcanic rhythm	火山韵律
volcanic rock	火山岩
volcanic sand	火山砂
volcanic structure	火山构造
volcanic type U-ore	火山岩型铀矿
volcanic vent	火山喷[火]口,火山通道
volcanism	火山作用
volcano	火山
volcanogenic massive sulfide deposit	火山成因块状硫化物矿床
volcanogenic mineral deposit	火山成因矿床
volcanogenic ore-forming process	火山成矿作用
volcanology	火山学
volcano-sedimentary boron deposit	火山沉积型硼矿床
volcano-sedimentary deposit	火山沉积型矿床
volcano-tectonic depression	火山-构造拗陷

VSMOW	VSMOW 标准(维也纳标准平均大洋水,国际原子能机构推荐的氢、氧同位素标准,δD=0)
VSMOW(Vienna standard mean ocean water)	VSMOW 标准(维也纳标准平均大洋水,国际原子能机构推荐的氢、氧同位素标准,δD=0)
vuggy	晶洞的,多孔的

W

wacke	瓦克岩,玄武质砂岩;玄武岩风化物,玄武土
wackestone	颗粒泥晶灰岩
Walker cell	沃克环流
wall rock	围岩
Wallis and Futuna Islands	瓦利斯群岛和富图纳群岛(法)
wallrock alteration	围岩蚀变
warm pool	暖池
Washington	华盛顿州(美)
washland	河漫滩
water drive reservoir	水驱储油层
water injection well	注水[油]井
water permeability	透水性
water pollution	水污染
water saturation	含水饱和度
water-rock interaction	水岩作用
wave crest line	波脊线
wave erosion	海蚀[作用]
wave refraction	波浪折射
wave-dominated delta	浪控三角洲
wavelength	褶皱波长
Wavelength Dispersive Spectroscopy(WDS)	波长色散光谱
Wavelength Dispersive X-ray Spectroscopy (WDXS)	波长分散 X 射线光谱

wavy cross-bedding	波状交错层理
wax，paraffin wax	石蜡
waxy luster	蜡状光泽
WCA	湿化学分析法
WDS	波长色散光谱
WDXS	波长分散 X 射线光谱
weather	天气
weathering	风化[作用]
weathering crust	风化壳
weathering crust ion-adsorbed REE deposit	风化壳离子吸附型稀土矿床
weathering crust mineral deposit	风化壳矿床
weathering zone	风化带
weathering-crust bauxite deposit	风化壳型铝土矿矿床
websterite	二辉岩
wedge cross-bedding	楔状交错层理
weishanite	围山矿
welded tuff	熔结凝灰岩
welded tuff texture	熔结凝灰结构
well rounded	极圆的
Wenlock	文洛克统
West Virginia	西弗吉尼亚州(美)
Western Sahara	西撒哈拉
Western Samoa	西萨摩亚
Wet Chemical Analysis(WCA)	湿化学分析法
wet gas	湿气
wet land	湿地
wheat	小麦[黄]色，淡黄色
white	白色
white dwarf	白矮星
white schist	白片岩
whitlockite	白磷钙石
Widmannstatten pattern	维德曼施泰滕相
willemite	硅锌矿

Wilson cycle	威尔逊旋回
wilting point	萎蔫点
winchite	蓝透闪石
wind direction	风向
wind erosion	风蚀
wind speed	风速
wind storm	风暴
wind-accumulated landform	风积地貌
wind-drift sand	风沙流
wind-dune cross-bedding	风成交错层理
wind-eroded castle	风城
wind-sand action	风沙作用
window	构造窗
wine	暗红色
wine red	葡萄酒红
winnowing	风选，风力选，簸选，吹蚀
Wisconsin	威斯康星州(美)
within-plate rock	板内岩石
Witwatersrand auri-uraniferous conglomerate	威特沃特斯兰金-铀砾岩
wolframite	黑钨矿(钨锰铁矿)
wolframite-quartz vein	黑钨矿-石英脉
wollastonite	硅灰石
wollastonite deposit	硅灰石矿床
wulfenite	钼铅矿，钼铅矿
Wulff net	吴氏网
wustite	方铁矿
Wyoming	怀俄明州(美)

X

XAFS	X 射线吸收精细结构光谱
XANES	X 射线吸收近边结构分析
Xenolith	重夕线石，捕虏体
xenomorphic granular	他形粒状

Xenon(Xe)	氙
xenothermal deposit	浅成高温水热矿床(浅成高温热液矿床)
xenotime	磷钇矿
xiangjiangite	湘江铀矿
xifengite	喜峰矿
xilingolite	锡林郭勒矿
ximengite	西盟石
xinganite, hingganite	兴安石
xitieshanite	锡铁山石
XLA-ICP-MS	X 射线激光磨蚀电感偶合质谱仪
XPES	光子发射光谱
XPS	X 射线光电子光谱
XRAS	X 射线吸收光谱
X-ray Absorption fine-Structure Spectroscopy (XAFS)	X 射线吸收精细结构光谱
X-ray Absorption Near-Edge Structure Analysis(XANES)	X 射线吸收近边结构分析
X-ray Absorption Spectroscopy (XRAS)	X 射线吸收光谱
X-ray crystallography	X 射线晶体学
X-ray Diffraction(XRD)	X 射线衍射
X-ray Fluorescence Spectroscopy(XRFS)	X 射线荧光谱
X-ray Laser Ablation Inductively Coupled Mass Spectroscopy(XLA-ICP-MS)	X 射线激光磨蚀电感偶合质谱仪
X-ray luminescence	X 射线发光
X-ray Luminescence(XRL)	X 射线发光分析
X-ray Photoelectron Spectroscopy(XPS)	X 射线光电子光谱
X-ray Powder Diffraction(XRPD)	X 射线粉晶衍射
X-ray Single Crystal Diffraction(XRSCD)	X 射线单晶衍射
X-ray Spectrometry(XRS)	X 射线光谱
XRD	X 射线衍射
XRFS	X 射线荧光谱

XRL	X 射线发光分析
XRPD	X 射线粉晶衍射
XRS	X 射线光谱
XRSCD	X 射线单晶衍射

Y

yellow	黄色
yellowish	带黄色的
yellowish white	黄白色
yellowness	黄色
yellowy	淡黄
Yemen	也门
yimengite	沂蒙矿
yoderite	紫硅镁铝石
Younger Dryas event	新仙女木事件
younging [direction]	面向
Ytterbium(Yb)	镱
Yttrium(Y)	钇
Yugoslavia	南斯拉夫
Yukon Territory	育空地区(加)

Z

Zabuyelite	扎布耶石
Zahl[德](或简称 Z)	单位晶胞"分子"数(或简称 Z 数)
Zaire	扎伊尔
Zambia	赞比亚
zeolite deposit	沸石矿床
zeolite facies	沸石相
zeolitic water	沸石水
zeolitization	沸石化
zeunerite	铜砷铀云母
Zhanghengite	张衡矿
Zimbabwe	津巴布韦
Zinc(Zn)	锌

zincobotryogen	锌赤铁矾
zincocopiapite	锌叶绿矾
zincovoltaite	锌氯钾铁矾
zinnwaldite	铁锂云母
zircon	锆[英]石
zirconium deposit	锆矿床
Zirconium(Zr)	锆
zoisite	黝帘石
zonal distribution of mineral deposits	矿床分带
zone	晶带
zone axis	晶带轴
zone index	晶带指数
zone law	晶带定律
zone of saturation	饱水带
zone of secondary enrichment	次生富集带
zone symbol	晶带符号
zoning sequence	分带序列
zoology	生态

3. 附录　Appendixes

附录 1　化学元素代号
Symbols of the Chemical Elements

代号 symbols	英文名 English Names	中文名 Chinese names
Ac	Actinium	锕
Al	Aluminium	铝
Am	Americium	镅
Sb	Antimony	锑
Ar	Argon	氩
As	Arsenic	砷
At	Astatine	砹
Ba	Barium	钡
Bk	Berkelium	锫
Be	Beryllium	铍
Bi	Bismuth	铋
B	Boron	硼
Br	Bromine	溴
Cd	Cadmium	镉
Cs	Caesium	铯
Ca	Calcium	钙
Cf	Californium	锎
C	Carbon	碳
Ce	Cerium	铈
Cl	Chlorine	氯
Cr	Chromium	铬
Co	Cobalt	钴

Cu	Copper	铜
Cm	Curium	锔
Dy	Dysprosium	镝
Es	Einsteinium	锿
Er	Erbium	铒
Eu	Europium	铕
Fm	Fermium	镄
F	Fluorine	氟
Fr	Francium	钫
Gd	Gadolinium	钆
Ga	Gallium	镓
Ge	Germanium	锗
Au	Gold	金
Hf	Hafnium	铪
He	Helium	氦
Ho	Holmium	钬
H	Hydrogen	氢
In	Indium	铟
I	Iodine	碘
Ir	Iridium	铱
Fe	Iron	铁
Kr	Krypton	氪
La	Lanthanum	镧
Lr	Lawrencium	铹
Pb	Lead	铅
Li	Lithium	锂
Lu	Lutetium	镥
Mg	Magnesium	镁
Mn	Manganese	锰
Md	Mendelevium	钔
Hg	Mercury	汞
Mo	Molybdenum	钼
Nd	Neodymium	钕

Ne	Neon	氖
Np	Neptunium	镎
Ni	Nickel	镍
Nb	Niobium	铌
N	Nitrogen	氮
No	Nobelium	锘
Os	Osmium	锇
O	Oxygen	氧
Pd	Palladium	钯
P	Phosphorus	磷
Pt	Platinum	铂
Pu	Plutonium	钚
Po	Polonium	钋
K	Potassium	钾
Pr	Praseodymium	镨
Pm	Promethium	钷
Pa	Protactinium	镤
Ra	Radium	镭
Rn	Radon	氡
Re	Rhenium	铼
Rh	Rhodium	铑
Rb	Rubidium	铷
Ru	Ruthenium	钌
Sm	Samarium	钐
Sc	Scandium	钪
Se	Selenium	硒
Si	Silicon	硅
Ag	Silver	银
Na	Sodium	钠
Sr	Strontium	锶
S	Sulphur	锍
Ta	Tantalum	钽
Tc	Technetium	锝

Te	Tellurium	碲
Tb	Terbium	铽
Tl	Thallium	铊
Th	Thorium	钍
Tm	Thulium	铥
Sn	Tin	锡
Ti	Titanium	钛
W	Tungsten	钨
U	Uranium	铀
V	Vanadium	钒
Xe	Xenon	氙
Yb	Ytterbium	镱
Y	Yttrium	钇
Zn	Zinc	锌
Zr	Zirconium	锆

附录 2　矿物代号　Mineral Symbols

代号 symbols	英文名 English Names	中文名 Chinese names
Ab	Albite	钠长石
Ac	Acmite	锥辉石
Act	Actinolite	阳起石
Ae	Aegirine	霓石
Af	Alkali feldspar	碱性长石
Alm	Almandine	铁铝榴石
Als	Al_2SiO_5 mineral	Al_2SiO_5 矿物
Am	Amphibole	角闪石
An	Anorthite	钙长石
And	Andalusite	红柱石
Andr	Andradite	钙铁榴石
Anl	Analcite	方沸石
Ann	Annite	羟铁云母
Ant	Anthophyllite	直闪石
Ar	Aragonite	文石
Aug	Augite	普通辉石
Bi	Biotite	黑云母
Brc	Brucite	水镁石
Cc	Calcite	方解石
Ch	Chlorite	绿泥石
Cld	Chloritoid	硬绿泥石
Co	Corundum	刚玉
Coe	Coesite	柯石英
Cpx	Clinopyroxene	单斜辉石
Crd	Cordierite	堇青石
Crs	Crossite	青铝闪石
Cum	Cummingtonite	镁铁闪石
Di	Diopside	透辉石
Dia	Diamond	金刚石

Dol	Dolomite	白云石
En	Enstatite	顽火辉石
Ep	Epidote	绿帘石
F	Fluid(H_2O)	流体(H_2O)
Fa	Fayalite	铁橄榄石
Fo	Forsterite	镁橄榄石
Fs	Ferrosilite	铁辉石
Ga	Garnet	石榴子石
Ged	Gedrite	铝直闪石
Gl	Glaucophane	蓝闪石
Gra	Graphite	石墨
Gro	Grossular	钙铝榴石
Gt	Garnet	石榴石
Hb	Hornblende	普通角闪石
Hd	Hedenbergite	钙铁辉石
Hu	Heulandite	片沸石
Hy	Hypersthene	紫苏辉石
Ill	Illite	伊利石
Jd	Jadeite	硬玉
Kao	Kaolinite	高岭石
Kf	K-feldspar	钾长石
Kp	Kaliophilite	钾霞石
Ky	Kyanite	蓝晶石
L	Liquid, melt	液体 = 熔体
Lc	Leucite	白榴石
Lm	Laumontite	浊沸石
Lw	Lawsonite	硬柱石
Mar	Margarite	珍珠云母
Mc	Mica	云母
Mgn	Magnesite	菱镁矿
Mi	Microcline	微斜长石
Mm	Montmorillonite	蒙脱石
Ms	Muscovite	白云母

Mt	Magnetite	磁铁矿
Ne	Nepheline	霞石
Ol	Olivine	橄榄石
Om	Omphacite	绿辉石
Opx	Orthopyroxene	斜方辉石
Or	Orthoclase	正长石
Pg	Paragonite	钠云母
Phl	Phlogopite	金云母
Phn	Phengite	多硅白云母
Pie	Piemontite	红帘石
Pig	Pigeonite	易变辉石
Prh	Prehnite	葡萄石
Prl	Pyrophyllite	叶蜡石
Pu	Pumpellyite	绿纤石
Px	Pyroxene	辉石
Pyr	Pyrope	镁铝榴石
Q	Quartz	石英
Rie	Riebeckite	钠闪石
San	Sanidine	透长石
Se	Sericite	绢云母
Ser	Serpentine	蛇纹石
Sil	Sillimanite	夕线石
Sm	Smectite	蒙皂石
Sp	Spinel	尖晶石
Sps	Spessartine	锰铝榴石
St	Staurolite	十字石
Stp	Stilpnomelane	黑硬绿泥石
Tc	Talc	滑石
Tr	Tremolite	透闪石
Ts	Tschermakite	契尔马克分子
V	Vapour	蒸汽
Ver	Vermiculite	蛭石
Wm	White mica	白色云母

Wo	Wollastonite	硅灰石
Wr	Wairakite	斜钙沸石
Wu	Wustite	方铁矿
Zo	Zoisite	黝帘石

附录 3　国际地层表

International Stratigraphic Chart

国际地层委员会 International Commission on Stratigraphy (2006)

Eonothem/宇 Eon /宙	Erathem/界 Era /代	System/系 Period /纪	Series/统 Epoch/世	Age/年代 (Ma)	Symbol/代号	
Phanerozoic PH 显生宇	Cenozoic CZ 新生界	Neogene 新近系	Holocene 全新统	0.011 8	Q_2	Q
			Pleistocene 更新统	1.806	Q_1	
			Pliocene 上新统	5.332	N_2	N
			Miocene 中新统	23.03	N_1	
		Paleogene 古近系	Oligocene 渐新统	33.9±0.1	E_3	E
			Eocene 始新统	55.8±0.2	E_2	
			Paleocene 古新统	65.5±0.3	E_1	
	Mesozoic MZ 中生界	Cretaceous 白垩系	Upper/Late 上白垩统	99.6±0.9	K_2	K
			Lower/Early 下白垩统	145.5±4.0	K_1	
		Jurassic 侏罗系	Upper/Late 上侏罗统	161.2±4.0	J_3	J
			Middle 中侏罗统	175.6±2.0	J_2	
			Lower/Early 下侏罗统	199.6±0.6	J_1	
		Triassic 三叠系	Upper/Late 上三叠统	228.0±2.0	T_3	T
			Middle 中三叠统	245.0±1.5	T_2	
			Lower/Early 下三叠统	251.0±0.4	T_1	

续表 1

Eonothem/宇 Eon /宙	Erathem/界 Era /代	System/系 Period /纪	Series/统 Epoch/世	Age/年代 (Ma)	Symbol/代号	
Phanerozoic PH 显生宇	Paleozoic PZ 古生界	Permian 二叠系	Lopingian 乐平统	260.4±0.7	P_3	P
			Guadalupian 瓜德鲁普统	270.6±0.7	P_2	
			Cisuralian 乌拉尔统	299.0±0.8	P_1	
		Carboniferous 石炭系	Pennsylvanian 宾夕法尼亚亚系	318.1±1.3	C_2	C
			Mississippian 密西西比亚系	359.2±2.5	C_1	
		Devonian 泥盆系	Upper/Late 上泥盆统	385.3±2.6	D_3	D
			Middle 中泥盆统	397.5±2.7	D_2	
			Lower/Early 下泥盆统	416.0±2.8	D_1	
		Silurian 志留系	Pridoli 普里多利统	418.7±2.7	S_4	S
			Ludlow 拉德洛统	422.9±2.5	S_3	
			Wenlock 文洛克统	428.2±2.3	S_2	
			Llandovery 兰多弗里统	443.7±1.5	S_1	
		Ordovician 奥陶系	Upper/Late 上奥陶统	460.9±1.6	O_3	O
			Middle 中奥陶统	471.8±1.6	O_2	
			Lower/Early 下奥陶统	488.3±1.7	O_1	
		Cambrian 寒武系	Furongian 芙蓉统	501.0±2.0	$\in_4$	∈
			Series 3 三统	510.0	$\in_3$	
			Series 2 二统	521.0	$\in_2$	
			Series 1 一统	542.0±1.0	$\in_1$	

续表 2

Eonothem/宇 Eon /宙	Erathem/界 Era /代	System/系 Period /纪	Series/统 Epoch/世	Age/年代 (Ma)	Symbol/ 代号	
Precambrian PC 前寒武系	Proterozoic PR 元古宇	Neoproterozoic 新元古界	Ediacaran 伊迪卡拉系	542 630	NP_3	NP
			Cryogenian 成冰系	850	NP_2	
			Tonian 拉伸系	1 000	NP_1	
		Mesoproterozoic 中元古界	Stenian 狭带系	1 200	MP_3	MP
			Ectasian 延展系	1 400	MP_2	
			Calymmian 盖层系	1 600	MP_1	
		Paleoproterozoic 古元古界	Statherian 固结系	1 800	PP_4	PP
			Orosirian 造山系	2 050	PP_3	
			Rhyacian 层侵系	2 300	PP_2	
			Siderian 成铁系	2 500	PP_1	
	Archean AR 太古宇	Neoarchean 新太古界	未再分系	2 800	NA	AR
		Mesoarchean 中太古界		3 200	MA	
		Paleoarchean 古太古界		3 600	PA	
		Eoarchean 始太古界			EA	

参考文献

References

《英汉地质词典》编辑组. 英汉地质词典. 北京:地质出版社,1983

地质矿产部地质辞典办公室. 地质辞典(二) 矿物岩石地球化学分册. 北京:地质出版社,1981

地质矿产部地质辞典办公室. 地质辞典(一) 普通地质构造地质分册. 北京:地质出版社,1983

地质矿产部地质辞典办公室. 地质辞典(二) 矿床地质应用地质分册. 北京:地质出版社,1986

韩吟文,马振东主编. 地球化学(面向 21 世纪课程教材). 北京:地质出版社,2003

黄定华. 普通地质学. 北京:高等教育出版社,2004

刘本培,蔡运龙. 地球科学导论. 北京:地质出版社,2000

刘心全,葛亚非,陈清江. 汉英常用地质学词汇. 武汉:中国地质大学出版社,1980

路凤香,桑隆康. 岩石学. 北京:地质出版社,2002

全国自然科学名词审定委员会. 地质学名词. 北京:科学出版社,1993

桑隆康. 变质岩岩石学术语汇编. 武汉:中国地质大学出版社,2001

武汉地质学院外语教研室. 英汉常用地质学词汇. 北京:科学出版社,1980

Best M G, Christiansen E H. Igneous Petrology. Oxford: Blackwell Scientific Publications, 2001

Blatt H, Trecy R J. Petrology. 2nd ed. New York: W H Freeman and Company,1996

Boggs S H Jr. Principles of Sedimentology and Stratigraphy. 2nd ed. NJ: Prentice-Hall Inc, Englewood Cliffs,1995

Cotter-Howells J D, Campbell L S, Valsami-Jones E, Batchelder M. The Mineralogical Society Series-9: Environmental Mineralogy. London: The Mineralogical Society of Great Britain and Ireland, 2000

Deer W A, Howie R A, Zussman J. Rock-Forming Minerals (Volume 4A): Framework Silicates——Feldspars. 2nd ed. London: Geological Society,

2001

Clare P Marshall, Rhodes W. Fairbridge, Encyclopaedia of geochemistry Dordrecht: Kluwer Academic Publishers, 1999. 712

Gary M, McAfee Jr R, Wolff C L. Glossary of Geology. Washington D C: American Geological Institute, 1973

Hibbard M J. Mineralogy—A Geologist's Point of View. New York: McGraw-Hill Higher Education, 2002

Hibbard M J. Petrography to Petrogenesis. New Jersey, U. S. A.: Prentice Hall, Englewood Cliffs, 1995

International Commission on Stratigraphy. International Stratigraphic Chart, 2006

Kearey P. The New Penguin Dictionary of Geology. 2nd ed. London: Penguin Books Ltd, 2001

Mason R. Petrology of Metamorphic Rocks. London: Geoge Allen & Univin Ltd, 1978, 1990

Mason R. Metamorphic Petrology. Wuhan: China University of Gosciences, 2000

Mehnert K R. Migmatites and the Origin of Granitic Rocks, New York: Elsevier Publishing Company, 1968

Miyashiro A. Metamorphic Petrology. London: UCL Press, 1994

Plummer C C, McGeary D, Carlson D. Physical Geology. 8th ed. New York: McGraw-Hill, 2001

Press F., Siever R. Earth. 4th ed. New York: W H Freeman and Company, 1986

Price E. Mathematical Techniques in Crystallography and Materials Science. 2nd ed. Berlin Heidelberg: Springer-Verlag, 1994

Price G D, Ross N L. The Mineralogical Society Series-3: The Stability of Minerals. London: Chapman & Hall, 1992

Raymond L A. Petrology: The study of Igneous, Sedimentary and Metamorphic Rocks. New York: Wm. C. Brown Communication Inc., 1995

Skinner B J, Porter S C. The Dynamic Earth. 2nd ed. New York: John Wiley & Sons, Inc, 1992